WORKED E

A LEVEL

PURE MATHS

by

C.E. Walsh

CELTIC REVISION AIDS

CELTIC REVISION AIDS
Lincoln Way, Windmill Road,
Sunbury on Thames, Middlesex

First published in this edition 1979

ISBN 017 751156 7

Printed in Hong Kong

PREFACE

This book is designed to help students reading mathematics for public examinations including GCE 'A' level and university students at the beginning of their courses.

One part of the book consists of a comprehensive range of worked examples and each section contains full explanations, where necessary, of any steps that the student may find difficult. Many of the examples, designated by (Ox), have been chosen from past Oxford University General Certificate of Education Advanced Level Examination papers. I am very grateful to the Oxford Delegacy of Local Examinations for permission to use these questions.

The answers in Part I should not be regarded as rigid models, but rather as illustrations of suitable logical solutions.

Part II of the Book contains notes on these worked examples. The notes have been expanded to include any general information which I think helpful for the solutions of other problems similar to those in Part I.

Consider examples 1 and 2. The same method will solve many other problems which may appear to be different or harder. This is explained in the notes. Similarly a general rule as illustrated in example 3a, and again in 3b, the possible applications of the mathematical device used are explained.

The variety of problems is endless. The following suggestions will be found helpful.

1. Study the syllabus thoroughly and carefully.
2. Become familiar with the bookwork needed. At first concentrate on material that you can understand easily and gradually progress to more difficult parts.
3. Try as many examples as possible. If you find some are too difficult, try others and then return to the unsolved ones.
4. When examining a problem decide the bookwork and principles involved.
5. Become familiar with the most common methods used in the solution of these problems and practice their applications.

C.E.W.

ACKNOWLEDGEMENTS

The publishers wish to thank the following for permission to use actual examinations questions in this book:

The Delegates of Local Examinations, University of Oxford.

PART I

1. (a) If: $$y = \frac{3 - x}{x^2 + 4x + 3}$$

find the value of:

$$\frac{d^n y}{dx^n}$$

where n is any positive integer.

(b) If: $$y = e^{ax} . \sin (bx + c)$$

find the value of:

$$\frac{d^n y}{dx^n}$$

(a) First express y in partial fractions:

$$y = \frac{2}{x + 1} - \frac{3}{x + 3}$$

Let: $$\frac{2}{x + 1} = u \ ; \ \frac{3}{x + 3} = v$$

then find: $\frac{d^n u}{dx^n}$ and $\frac{d^n v}{dx^n}$ separately.

$$\frac{du}{dx} = 2 . \frac{(-1)}{(x + 1)^2} = (-1) . 2 . (x + 1)^{-2}$$

$$\frac{d^2 u}{dx^2} = (-1) . 2 . (-2) . (x + 1)^{-3}$$

$$= (-1)^2 . 2 . 2!(x+1)^{-3}$$

$$\frac{d^3u}{dx^3} = (-1)^2 . 2! 2 . (-3) . (x+1)^{-4}$$

$$= (-1)^3 . 2 . 3!(x+1)^{-4}$$

and so, generally:

$$\frac{d^nu}{dx^n} = (-1)^n . \frac{n! 2}{(x+1)^{n+1}}$$

Similarly:

$$\frac{d^nv}{dx^n} = (-1)^n . \frac{n . 3}{(x+3)^{n+1}}$$

Therefore:

$$\frac{d^ny}{dx^n} = \frac{d^nu}{dx^n} - \frac{d^nv}{dx^n}$$

$$= (-1)^n . n! \left(\frac{2}{(x+1)^{n+1}} - \frac{3}{(x+3)^{n+1}}\right)$$

(b) Differentiating:

$$\frac{dy}{dx} = e^{ax} . b \cos(bx+c) + a . e^{ax} \sin(bx+c)$$

$$= e^{ax}(b \cos(bx+c) + a \sin(bx+c))$$

Now let $a = r \cos \alpha$, $b = r \sin \alpha$; so that, in the usual way:

$$r = (a^2 + b^2)^{\frac{1}{2}} \quad ; \quad \text{and: } \alpha = \tan^{-1}\frac{b}{a}$$

Then $b \cos(bx+c) + a \sin(bx+c)$ becomes:

$$r(\sin \alpha \cos (bx + c) + \cos \alpha \sin (bx + c))$$

$$= r \sin (bx + c + \alpha)$$

Hence:

$$\frac{dy}{dx} = r \cdot e^{ax} \cdot \sin (bx + c + \alpha)$$

Similarly, differentiating again:

$$\frac{d^2y}{dx^2} = r^2 \cdot e^{ax} \cdot \sin (bx + c + \alpha + \alpha)$$

$$= r^2 \cdot e^{ax} \cdot \sin (bx + c + 2\alpha)$$

and so on. Thus, generally:

$$\frac{d^ny}{dx^n} = r^n \cdot e^{ax} \cdot \sin (bx + c + n\alpha)$$

where:

$$r = (a^2 + b^2)^{\frac{1}{2}} \quad ; \quad \alpha = \tan^{-1} \frac{b}{a}$$

2. Prove that:

$$\frac{d}{dt}\left(\frac{t^2}{\sin t}\right) > 0 \text{ for } 0 < t < \tfrac{1}{2}\pi$$

Prove that:

$$\int_0^{\pi/2} \{1 - \cos (2n + 1)t\} \frac{d}{dt}\left(\frac{t^2}{\sin t}\right) dt > 0$$

and that:

$$\int_0^{\pi/2} \cos(2n+1)t \cdot \frac{d}{dt}\left(\frac{t^2}{\sin t}\right) dt < \tfrac{1}{4}\pi^2$$

By integrating by parts, prove that, when n is a positive integer:

$$\int_0^{\pi/2} \left(\frac{t^2}{\sin t}\right) \cdot \sin(2n+1)t \cdot dt < \frac{\pi^2}{4(2n+1)}$$

(Ox)

First:

$$\frac{d}{dt}\left(\frac{t^2}{\sin t}\right) = \frac{\sin t \cdot 2t - t^2 \cos t}{(\sin t)^2}$$

$$= \frac{t \cos t(2 \tan t - t)}{(\sin t)^2}$$

Since: $0 < t < \tfrac{1}{2}\pi$; $\cos t > 0$ and $\tan t > t$

Hence also: $2 \tan t > t$

Thus: $\dfrac{t \cos t\,(2 \tan t - t)}{(\sin t)^2} > 0$

i.e: $\dfrac{d}{dt}\left(\dfrac{t^2}{\sin t}\right) > 0$

also since: $1 - \cos(2n+1)t \geqslant 0$ it follows

that: $$\int_0^{\pi/2} \left\{1 - \cos(2n+1)t\right\} \frac{d}{dt}\left(\frac{t^2}{\sin t}\right) dt > 0$$

Separating this into two integrals, we see that:

$$\int_0^{\pi/2} \frac{d}{dt}\left(\frac{t^2}{\sin t}\right)dt > \int_0^{\pi/2} \cos(2n+1)t \cdot \frac{d}{dt}\left(\frac{t^2}{\sin t}\right)dt \qquad (1)$$

now:

$$\int_0^{\pi/2} \frac{d}{dt}\left(\frac{t^2}{\sin t}\right)dt = \left[\frac{t^2}{\sin t}\right]_0^{\pi/2} = \frac{\pi^2}{4} - 0 = \frac{\pi^2}{4}$$

Therefore, from (1):

$$\int_0^{\pi/2} \cos(2n+1)t \cdot \frac{d}{dt}\left(\frac{t^2}{\sin t}\right)dt < \frac{\pi^2}{4}$$

Finally, integrating by parts:

$$\int_0^{\pi/2} \frac{t^2}{\sin t} \cdot \sin(2n+1)t \cdot dt = \left[-\frac{t^2}{\sin t} \cdot \frac{\cos(2n+1)t}{2n+1}\right]_0^{\pi/2}$$

$$+ \int_0^{\pi/2} \frac{\cos(2n+1)t}{2n+1} \cdot \frac{d}{dt}\left(\frac{t^2}{\sin t}\right)dt$$

When n is a positive integer:

$$\cos(2n+1)\pi/2 = 0$$

so:

$$\left[-\frac{t^2}{\sin t} \cdot \frac{\cos(2n+1)t}{2n+1}\right]_0^{\pi/2} = 0$$

Thus:

$$\int_0^{\pi/2} \frac{t^2}{\sin t} \cdot \sin (2n + 1)t \cdot dt =$$

$$\frac{1}{2n + 1} \cdot \int_0^{\pi/2} \cos (2n + 1)t \cdot \frac{d}{dt}\left(\frac{t^2}{\sin t}\right)dt < \frac{\pi^2}{4(2n + 1)}$$

3. Find:

(a) $\int \frac{(x + 4)dx}{\sqrt{(2x + 3)}}$

(b) $\int \frac{x^3 dx}{(x^2 + a^2)^2}$

(c) $\int \cos^2 \theta \cos 3\theta \cdot d\theta$

(a) Let $2x + 3 = u^2$, so that $2dx$ is replaced by $2udu$, i.e. dx by udu, x by $(u^2 - 3)/2$. The integral now becomes:

$$\int \frac{(u^2 + 5)}{2u} udu = \tfrac{1}{2} \int (u^2 + 5)du = \tfrac{1}{2}\left(\frac{u^3}{3} + 5u\right)$$

Replacing u by $\sqrt{(2x + 3)}$:

$$= \frac{\sqrt{(2x + 3)}(x + 9)}{3}$$

on reduction.

(b) The integrand can be written in the form:

$$\frac{x^2 \quad xdx}{\left(x^2 + a^2\right)^2}$$

This suggests the substitution $u = x^2 + a^2$, so that xdx is replaced by du/2. The integral transforms into:

$$\tfrac{1}{2}\int \frac{(u - a^2)}{u^2}du = \tfrac{1}{2}\left[\int \frac{du}{u} - a^2\int \frac{du}{u^2}\right] = \tfrac{1}{2}\left(\log u + \frac{a^2}{u}\right)$$

$$= \tfrac{1}{2}\left(\log (a^2 + x^2) + \frac{a^2}{a^2 + x^2}\right)$$

(c) Replace $\cos^2 \theta$ by:

$$\frac{1 + \cos 2\theta}{2}$$

so that the integral becomes:

$$\tfrac{1}{2}\int (1 + \cos 2\theta) \cos 3\theta \, . \, d\theta = \tfrac{1}{2}\left(\int \cos 3\theta \, d\theta + \int \cos 2\theta \, . \cos 3\theta \, . \, d\theta\right) \quad (1)$$

now: $$\int \cos k\theta \, . \, d\theta = \frac{\sin k\theta}{k}$$

so that: $$\int \cos 3\theta \, . \, d\theta = \frac{\sin 3\theta}{3}$$

also:

$$\int \cos 2\theta \, . \cos 3\theta \, . \, d\theta = \tfrac{1}{2}\int (\cos 5\theta + \cos \theta)d\theta$$

$$= \tfrac{1}{2}\left(\frac{\sin 5\theta}{5} + \sin \theta\right)$$

Therefore substitution in (1) gives:

$$\tfrac{1}{2}\left[\frac{\sin 3\theta}{3} + \tfrac{1}{2}\left(\frac{\sin 5\theta}{5} + \sin \theta\right)\right]$$

so, rearranging:

$$\int \cos^2 \theta \cos 3\theta \, d\theta = \frac{\sin \theta}{4} + \frac{\sin 3\theta}{6} + \frac{\sin 5\theta}{20}$$

4. At each point on a given curve:

$$\frac{d^2y}{dx^2} = 2(3x + 5)$$

The gradient of the curve at the point where it cuts the y axis is -2. The curve passes through the point (-1, 4). Find its equation and point of inflexion.

$$\frac{d^2y}{dx^2} = 2(3x + 5)$$

Integrate with respect to x:

$$\frac{dy}{dx} = 3x^2 + 10x + A$$

where A is a constant to be found.

From the given conditions:

$$\frac{dy}{dx} = -2$$

when x = 0, therefore A = -2. And so:

$$\frac{dy}{dx} = 3x^2 + 10x - 2$$

Integrating again:

$$y = x^3 + 5x^2 - 2x + B$$

where B is another constant to be found from the given conditions.

As the curve passes through the point (-1, 4), this gives B = -2, and the equation of the curve to be:

$$y = x^3 + 5x^2 - 2x - 2 \qquad (1)$$

To find the point of inflexion of the curve consider:

$$\frac{d^2y}{dx^2} = 0, \text{ i.e. } 2(3x + 5) = 0$$

This occurs when:

$$x = \frac{-5}{3}$$

also, at that point:

$$\frac{d^2y}{dx^2}$$

changes sign. Thus there is a point of inflexion at the point:

$$x = \frac{-5}{3}$$

on the curve, the corresponding value of y being calculated from (1) with $x = -5/3$. It is not necessary to do so here.

5. If: $$\frac{dy}{dx} - 3z = 6$$

and: $$\frac{dz}{dx} - 3y = 12$$

and: $$y = z = 0$$

when $x = 0$, find y and z in terms of x, and show that:

$$y^2 - z^2 + 8y - 4z = 0 \qquad \text{(Ox)}$$

Adding the two given equations:

$$\frac{d(y + z)}{dx} - 3(y + z) = 18 \qquad (1)$$

Subtracting them instead:

$$\frac{d(y - z)}{dx} + 3(y - 3) = -6 \qquad (2)$$

Let $y + z = u$, then (1) becomes:

$$\frac{du}{dx} = 3(u + 6)$$

or:
$$\frac{du}{u + 6} = 3dx$$

Integrating both sides, $\log(u + 6) = 3x + A$; A is a constant. To find A, from the given conditions, $u = y + z = 0$ when $x = 0$, and so $A = \log 6$. Thus $\log(y + z + 6) = 3x + \log 6$ or:

$$y + z + 6 = 6 \cdot e^{3x} \qquad (3)$$

Similarly replacing $y - z$ by v in (2) it becomes:

$$\frac{dv}{dx} = -3(v + 2)$$

or:
$$\frac{dv}{v + 2} = -3dx$$

Integration gives:

$$\log(v + 2) = -3x + B$$

where B is a constant.

When $x = 0$, $v = 0$, therefore $B = \log 2$. Hence:

$$\log(y - z + 2) = -3x + \log 2$$

i.e.
$$y - z + 2 = 2e^{-3x} \qquad (4)$$

Adding and subtracting (3) and (4) in turn:

$$y = 3e^{3x} + e^{-3x} - 4$$

$$z = 3e^{3x} - e^{-3x} - 2$$

To eliminate x, multiply (3) and (4), getting:

$$(y + z + 6)(y - z + 2) = 12$$

or, after multiplication:

$$y^2 - z^2 + 8y - 4z = 0$$

6. (a) Find the equations of the tangent and normal to the curve $y^3 + 2x^3 = 5xy$ at the point (1, 2) on it.

(b) Find the acute angle at which the curves $y^2 = 4x$ and $9x^2 - 2y^2 = 1$ intersect.

(a) Differentiate with respect to x:

$$3y^2 \cdot \frac{dy}{dx} + 6x^2 = 5y + 5x\frac{dy}{dx}$$

$$\therefore \quad \frac{dy}{dx} = \frac{5y - 6x^2}{3y^2 - 5x}$$

at the point (1, 2) this has the value 4/7. The equation of the tangent to the curve at this point will, therefore, be:

$$\frac{y - 2}{x - 1} = \frac{4}{7}$$

or: $4x - 7y + 10 = 0$

The equation of the normal will be:

$$\frac{y - 2}{x - 1} = \frac{-7}{4}$$

or: $7x + 4y - 15 = 0$

(b) Solving for x and y from the equations of the curves, they are seen to intersect at the points (1, 2) and (1 - 2).

To find the slopes of the curves at these points, differentiate the equation of each curve with respect to x.

For: $y^2 = 4x \qquad \frac{dy}{dx} = \frac{2}{y}$

At the point (1, 2) this has the value 1.

For: $9x^2 - 2y^2 = 1 \qquad \frac{dy}{dx} = \frac{9x}{2y}$

At the point (1, 2) this has the value 9/4.

Thus the tangents to the two curves at the point (1, 2) have slopes $m_1 = 1$, and $m_2 = 9/4$. Let θ be the acute angle at which the curves intersect, i.e. at which the tangents at (1, 2) intersect. Then:

$$\tan\theta = \frac{m_2 - m_1}{1 + m_2 m_1} = \frac{5}{13}$$

so: $$\theta = \tan^{-1}\frac{5}{13}$$

Similarly the acute angle between the curves at their other point of intersection is also:

$$\tan^{-1}\frac{5}{13}$$

7. Show that for each a > 0, the function:

$$e^{-ax} \cdot x^{a^2}$$

has a maximum value F(a) say, and that F(x) has a minimum value:

$$e^{-\frac{1}{2}e}$$

Show that for each b $>$ 1, the function:

$$e^{-xb} \cdot b^{x^2}$$

has a minimum value G(b), and that G(x) has a maximum value which is also:

$$e^{-\frac{1}{2}e}$$

(Ox)

Let: $u = e^{-ax} \cdot x^{a^2}$

then: $\frac{du}{dx} = e^{-ax}(a^2 \cdot x^{a^2-1} - a \cdot x^{a^2})$

$$= a : e^{-ax} \cdot x^{a^2-1}(a - x) = 0$$

when x = a.

Also, since a $>$ 0, du/dx changes from being positive to negative at x = a, so u has a maximum value when x = a, which is:

$$e^{-a^2} \cdot a^{a^2} = F(a)$$

And so:

$$F(x) = e^{-x^2} \cdot x^{x^2} = e^{-x^2} \cdot e^{x^2 \log x}$$

$$= e^{x^2 \log x - x^2}$$

Therefore:

$$\frac{dF(x)}{dx} = e^{x^2 \log x - x^2}(x^2 \cdot \frac{1}{x} + 2x \log x - 2x)$$

$$= e^{x^2 \log x - x^2} . (2x \log x - x)$$

$$= e^{x^2 \log x - x^2} . x(2 \log x - 1) = 0$$

when $\log x = \frac{1}{2}$.

Also for this value of x,

$$\frac{dF(x)}{dx}$$

changes from being negative to positive, so F(x) has a minimum value when $\log x = \frac{1}{2}$, i.e. when $x = e^{\frac{1}{2}}$. Substitution in F(x) shows this minimum value to be:

$$e^{\frac{1}{2}e - e} = e^{-\frac{1}{2}e}$$

Now let: $v = e^{-xb} . b^{x^2} = e^{-xb + x^2 \log b}$

Differentiating:

$$\frac{dv}{dx} = e^{-xb + x^2 \log b} . (-b + 2x \log b)$$

$$= 0, \text{ when } x = \frac{b}{2 \log b}$$

Since $b > 1$ this is positive.

Also $\frac{dv}{dx}$ changes from being negative to positive here,

so: $x = \frac{b}{2 \log b}$ gives a minimum value for v.

This minimum value is, on substituting for x:

$$e^{-\frac{b^2}{2 \log b} + \frac{b^2}{4 \log b}} = e^{-\frac{b^2}{4 \log b}} = G(b)$$

so that:

$$G(x) = e^{-\frac{x^2}{4 \log x}}$$

$$\frac{dG(x)}{dx} = e^{-\frac{x^2}{4 \log x}} \cdot \left(\frac{\frac{4x^2}{x} - 8x \log x}{(4 \log x)^2} \right) = 0$$

when $\log x = \frac{1}{2}$, i.e. when $x = e^{\frac{1}{2}}$.

As well, $\frac{dG(x)}{dx}$ changes from being positive to negative

here, so: $$x = e^{\frac{1}{2}}$$

gives a maximum value for G(x). Substitution shows this to be:

$$e^{-\frac{1}{2}e}$$

also.

8. Let:

$$I_{m,n} \equiv \int x^m \cdot (\log x)^n \cdot dx$$

Prove the reduction formula:

$$(m+1)I_{m,n} = x^{m+1} \cdot (\log x)^n - n \cdot I_{m,n-1}$$

Hence find: $$\int x^2 \cdot (\log x)^3 \cdot dx$$

Integrating by parts:

$$I_{m.n} = \frac{x^{m+1}}{m+1}(\log x)^n - \int \frac{x^{m+1}}{m+1} \cdot n(\log x)^{n-1} \cdot \frac{1}{x} dx$$

$$= \frac{x^{m+1}}{m+1}(\log x)^n - \frac{n}{m+1} \int x^m (\log x)^{n-1} \cdot dx$$

i.e.

$$I_{m,n} = \frac{x^{m+1}}{m+1}(\log x)^n - \frac{n}{m+1} I_{m,n-1} \qquad (1)$$

or:

$$(m+1)\, I_{m.n} = x^{m+1} . (\log x)^n - n . I_{m.n-1}$$

Put m = 2, n = 3 in (1) and it becomes:

$$\int x^2 (\log x)^3 dx = I_{2,3} = \frac{x^3 . (\log x)^3}{3} - I_{2,2} \qquad (2)$$

Similarly, with m = 2, n = 2:

$$I_{2,2} = \frac{x^3 . (\log x)^2}{3} - \frac{2}{3} I_{2,1} \qquad (3)$$

and:

$$I_{2,1} = \int x^2 \log x = \frac{x^3 \log x}{3} - \frac{x^3}{3^2} \qquad (4)$$

Therefore, from (2), (3) and (4):

$$I_{2,3} = \frac{x^3(\log x)^3}{3} - \left(\frac{x^3(\log x)^2}{3} - \frac{2}{3} I_{2,1}\right)$$

$$= \frac{x^3(\log x)^3}{3} - \frac{x^3(\log x)^2}{3} +$$

$$\frac{2}{3^2} . x^3 \log x - \frac{2x^3}{3^3}$$

9. Find:

(a) $\int x \,.\, \cos^2 x \,.\, dx$

(b) $\int \sqrt{\left(\frac{k - x}{k + x}\right)} .\, dx$

(a) Replace $\cos^2 x$ by:

$$\frac{1 + \cos 2x}{2}$$

then:

$$\int x \cos^2 x \, dx = \frac{1}{2}\int x(1 + \cos 2x)dx$$

$$= \frac{x^2}{4} + \frac{1}{2}\int x \cos 2x \,.\, dx \quad (1)$$

To find $\int x \cos 2x \,.\, dx$, replace $2x$ by u and it becomes:

$$\frac{1}{2}\int u \cos u \,.\, \frac{du}{2} = \frac{1}{4}\int u \cos u \,.\, du$$

Integrating by parts, this becomes:

$$\frac{1}{4}(u \sin u - \int \sin u \,.\, du)$$

$$= \frac{1}{4}(u \sin u + \cos u)$$

$$= \frac{1}{4}(2x \sin 2x + \cos 2x)$$

Substituting in (1):

$$\int x \cos^2 x \,.\, dx = \frac{x^2}{4} + \frac{1}{8}(2x \sin 2x + \cos 2x)$$

$$= \frac{1}{4}\left(x^2 + x \sin 2x + \frac{\cos 2x}{2}\right)$$

(b) $\int \sqrt{\left(\frac{k - x}{k + x}\right)}\ dx$

Rationalize the numerator and the integral becomes:

$$\int \frac{(k - x)}{\sqrt{((k + x)(k - x))}} dx = \int \frac{(k - x)}{\sqrt{\left(k^2 \quad x^2\right)}} dx$$

$$= k \int \frac{dx}{\sqrt{\left(k^2 - x^2\right)}} - \int \frac{xdx}{\sqrt{\left(k^2 - x^2\right)}}$$

$$= k \sin^{-1}\left(\frac{x}{k}\right) - \int \frac{xdx}{\sqrt{\left(k^2 - x^2\right)}}$$

If we let $k^2 - x^2 = u$, the last integral transforms into:

$$\int -\frac{1}{2} \cdot u^{-\frac{1}{2}} du = -\sqrt{\left(k^2 - x^2\right)}$$

Therefore:

$$\int \sqrt{\left(\frac{k - x}{k + x}\right)} dx = k \sin^{-1}\left(\frac{x}{k}\right) + \sqrt{\left(k^2 - x^2\right)}$$

10. (a) OX and OY are perpendicular axes of co-ordinates. A rectangle ABCD is drawn with A on OX, B on OY and CD on the opposite side of AB to O. A point P, whose co-ordinates are (h, k) lies on the opposite side of CD to O. Prove that the length of the perpendicular from P to CD in terms of angle OBA (= α), AB (= a), and BC (= b) is:

$$h \cos \alpha + k \sin \alpha - (a \sin \alpha \cos \alpha + b)$$

(b) The diagram is a plan of a right-angled turn in a corridor, the two arms of which are of widths h and k. A rectangular box, whose base is of length:

$$\sqrt{(h^2 + k^2)}$$

and breadth b, has to be pushed along the floor on its base round the corner in the corridor. Prove that this is possible only if:

$$b \leqslant \frac{hk}{\sqrt{(h^2 + k^2)}}$$

(Ox)

(a)

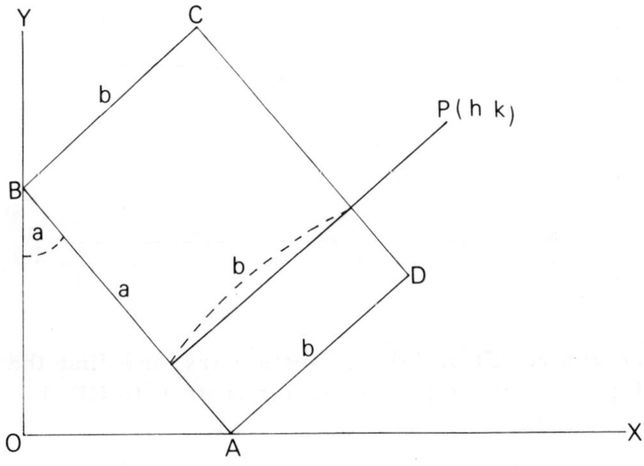

The equation of AB is easily seen to be:

$$\frac{x}{\sin a} + \frac{y}{\cos a} = a$$

or: $x \cos a + y \sin a - a \sin a . \cos a = 0.$

The length of PN, the perpendicular to AB from P (h, k) is therefore:

$$h \cos \alpha + k \sin \alpha - a \sin \alpha \cos \alpha$$

and so the length of the perpendicular from P to CD $= PN - b = h \cos \alpha + k \sin \alpha - (a \sin \alpha \cos \alpha + b)$.

(b)

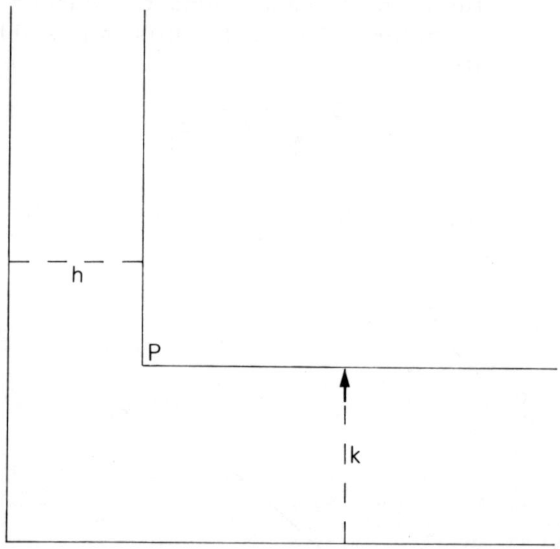

We use the result of (a). Let α vary and find the limiting value of the perpendicular from P to CD, i.e. of:

$$h \cos \alpha + k \sin \alpha - (a \sin \alpha \cos \alpha + b)$$

This must be positive or, in limiting case, zero. Let:

$$u = h \cos \alpha + k \sin \alpha - a \sin \alpha . \cos \alpha$$

$$\frac{du}{d\alpha} = -h \sin \alpha + k \cos \alpha - a(\cos^2 \alpha - \sin^2 \alpha)$$

Here $a = \sqrt{(h^2 + k^2)}$ so we let $h = a \sin \beta$, $k = a \cos \beta$.

$$\frac{du}{d\alpha} \text{ is now:}$$

$$a \cos \beta \cos \alpha - a \sin \beta \sin \alpha - a \cos 2\alpha$$

$$= a(\cos(\alpha + \beta) - \cos 2\alpha) = 0 \text{ when } \alpha = \beta$$

This is easily seen to give a minimum value for u. Thus minimum value of u is, when $\alpha = \beta$:

$$\frac{hk}{\sqrt{(h^2+k^2)}} + \frac{k \cdot h}{\sqrt{(h^2+k^2)}} - \sqrt{(h^2+k^2)} \cdot \frac{h}{\sqrt{(h^2+k^2)}} \cdot \frac{k}{\sqrt{(h^2+k^2)}}$$

$$= \frac{hk}{\sqrt{(h^2+k^2)}}$$

In order that the box may be pushed round the corner, from (a):

$$u - b \geqslant 0 \quad \text{always}$$

Therefore: $$\frac{hk}{\sqrt{(h^2+k^2)}} \geqslant b$$

11. Investigate the maxima and minima of the functions:

    (a) $x^4 - 4x^3 - 2x^2 + 12x + 1$

    (b) $4x^3 - 6x^2 + 3x + 2$

(a) Let $y = x^4 - 4x^3 - 2x^2 + 12x + 1$.

Then:

$$\frac{dy}{dx} = 4(x^3 - 3x^2 - x + 3) = 4(x - 3)(x + 1)(x - 1)$$

and so:

$$\frac{dy}{dx} = 0, \text{ when } x = -1, \; x = +1, \; x = +3$$

At these points the respective changes in sign of:

$$\frac{dy}{dx}$$

are from negative to positive, positive to negative, negative to positive. Thus y has a:

minimum value -8 when $x = -1$

maximum value +8 when $x = +1$

minimum value -8 when $x = +3$

(b) Let $y = 4x^3 - 6x^2 + 3x + 2$.

Then:

$$\frac{dy}{dx} = 12x^2 - 12x + 3 = 3(2x - 1)^2 = 0, \text{ when } x = \tfrac{1}{2}$$

However dy/dx does **not** change sign, as $3(2x - 1)^2$ is always positive except when $x = \frac{1}{2}$.

Thus $x = \frac{1}{2}$ does not give a maximum or minimum value for x.

Differentiating: $\dfrac{dy}{dx}$ we get:

$$\frac{d^2y}{dx^2} = 12(2x - 1) \quad \text{and so:}$$

$$\frac{d^2y}{dx^2} = 0 \text{ when } x = \tfrac{1}{2}$$

and changes sign there.

Therefore y has no maxima or minima, but a point of inflexion when $x = \frac{1}{2}$.

12. Prove that:

$$\int_0^1 |x - t| \cos \pi t \,.\, dt = \begin{cases} -\dfrac{2}{\pi^2} \cos \pi x \; (0 < x < 1) \\ \dfrac{2}{\pi^2} \; (x \geqslant 1) \\ -\dfrac{2}{\pi^2} \; (x \leqslant 0) \end{cases}$$

Evaluate similarly:

$$\int_0^1 |x - t| \sin \pi t \,.\, dt$$

(Ox)

If: $x \geqslant 1$, $|x - t| = x - t$ for $0 \leqslant t \leqslant 1$

and so:

$$\int_0^1 |x - t| \cos \pi t \,.\, dt = \int_0^1 (x - t) \cos \pi t \,.\, dt$$

integrating by parts:

$$= \left[(x - t) \frac{\sin \pi t}{\pi} - \frac{\cos \pi t}{\pi^2} \right]_0^1$$

$$= \frac{1}{\pi^2} - (-\frac{1}{\pi^2}) = \frac{2}{\pi^2}$$

If $x \leqslant 0$, $|x - t| = t - x$, and so:

$$\int_0^1 |x - t| \cos \pi t \,.\, dt$$

$$= -\int_0^1 (x - t) \cos \pi t \, . \, dt = -\frac{2}{\pi^2}$$

Now for the case $0 < x < 1$. Here:

$$\int_0^1 |x - t| \cos \pi t \, . \, dt = \int_x^1 (t - x) \cos \pi t \, . \, dt +$$

$$\int_0^x (x - t) \cos \pi t \, . \, dt$$

$$= \left[(t - x)\frac{\sin \pi t}{\pi} + \frac{\cos \pi t}{\pi^2} \right]_x^1 +$$

$$\left[(x - t)\frac{\sin \pi t}{\pi} - \frac{\cos \pi t}{\pi^2} \right]_0^x$$

$$= \left(-\frac{1}{\pi^2} - \frac{\cos \pi x}{\pi^2} \right) + \left(-\frac{\cos \pi x}{\pi^2} - \left(-\frac{1}{\pi^2} \right) \right)$$

$$= -\frac{2 \cos \pi x}{\pi^2}$$

Similarly if $x \geqslant 1$:

$$\int_0^1 |x - t| \sin \pi t \, . \, dt = \int_0^1 (x - t) \sin \pi t \, . \, dt$$

$$= \left[-(x - t)\frac{\cos \pi t}{\pi} - \frac{\sin \pi t}{\pi^2} \right]_0^1$$

$$= \frac{x - 1}{\pi} - \left(-\frac{x}{\pi} \right) = \frac{2x - 1}{\pi}$$

If $x \leqslant 0$:

$$\int_0^1 |x - t| \sin \pi t \, . \, dt = - \int_0^1 (x - t) \sin \pi t \, . \, dt = \frac{1 - 2x}{\pi}$$

Finally, if $0 < x < 1$:

$$\int_0^1 |x - t| \sin \pi t \, . \, dt = \int_x^1 (t - x) \sin \pi t \, . \, dt +$$

$$\int_0^x (x - t) \sin \pi t \, . \, dt$$

$$= \left[-(t - x) \frac{\cos \pi t}{\pi} + \frac{\sin \pi t}{\pi^2} \right]_x^1 -$$

$$\left[-(t - x) \frac{\cos \pi t}{\pi} + \frac{\sin \pi t}{\pi^2} \right]_0^x$$

$$= \frac{1 - x}{\pi} - \frac{\sin \pi x}{\pi^2} - \left(\frac{\sin \pi x}{\pi^2} - \frac{x}{\pi} \right)$$

$$= \frac{1}{\pi} - \frac{2 \sin \pi x}{\pi^2}$$

13. Evaluate:

(a) $\int_1^e x^{-5} . \log_e x . dx$

(b) $\int_0^1 e^{x^5} \quad x^4 dx$

(c) $\int_0^1 \frac{\sqrt{(x)} . dx}{\sqrt{(x)} + \sqrt{(1 - x)}}$

(a) Integrating by parts:

$$\int_1^e x^{-5} . \log x . dx = \left[\frac{-x^{-4}}{4}\log x\right]_1^e + \frac{1}{4}\int_1^e x^{-4} . \frac{1}{x} . dx$$

$$= \frac{-e^{-4}}{4} + \frac{1}{4}\int_1^e x^{-5} dx$$

$$= \frac{-e^{-4}}{4} + \frac{1}{4}\left(\frac{-e^{-4}}{4} + \frac{1}{4}\right) = \frac{1}{16}\left(1 - 5e^{-4}\right)$$

(b) Let $x^5 = u$ so that $5x^4dx$ is replaced by du, and:

$$\int_0^1 e^{x^5} . x^4 dx$$

becomes:

$$\int_0^1 e^u . \frac{du}{5} = \frac{1}{5}[e^u]_0^1 = \frac{1}{5}(e - 1)$$

(c) Use formula:

$$\int_0^a f(x) \, . \, dx = \int_0^a f(a - x)dx$$

Thus if:

$$I = \int_0^1 \frac{\sqrt{(x)} \, . \, dx}{\sqrt{(x)} + \sqrt{(1 - x)}}$$

replacing x by 1 - x:

$$I = \int_0^1 \frac{\sqrt{(1 - x)} \, . \, dx}{\sqrt{(1 - x)} + \sqrt{(x)}}$$

adding:

$$2I = \int_0^1 \frac{\sqrt{(x)} + \sqrt{(1 - x)}}{\sqrt{(x)} + \sqrt{(1 - x)}} \, . \, dx = \int_0^1 dx = 1$$

Hence: $I = \frac{1}{2}$

14. (a) Find the centroid of the area bounded by the parabola $y^2 = 4x$, the lines $x = 1$, $x = 4$, and the x axis.

(b) The area under the parabola $y^2 = 4ax$ between $x = 0$, and $x = b$ is rotated once about the x axis. Find the centroid of the solid generated.

(a)

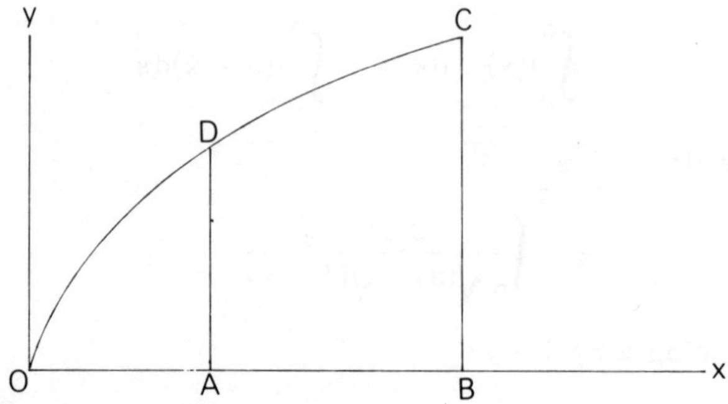

Let A and B be the points (1, 0), (4, 0) respectively, AD and BC the ordinates at A and B, so that ABCD is the area whose centroid is to be found. Let it be the point (X, Y), then by the usual formulae:

$$X = \frac{\int_1^4 xydx}{\int_1^4 ydx}$$

$$Y = \frac{\int_1^4 \frac{y^2}{2} dx}{\int_1^4 ydx}$$

at points on the parabola:

$$y^2 = 4x, \text{ or } y = 2\sqrt{(x)}$$

Substituting for y:

$$X = \frac{\int_1^4 x^{3/2} dx}{\int_1^4 x^{1/2} dx} \qquad \text{and:} \quad Y = \frac{\int_1^4 2xdx}{\int_1^4 2x^{1/2} dx} \qquad (1)$$

Now:

$$\int_1^4 x^{3/2}dx = \frac{2}{5}(4^{5/2} - 1) = \frac{62}{5}$$

$$\int_1^4 x^{1/2}dx = \frac{2}{3}(4^{3/2} - 1) = \frac{14}{3}$$

$$\int_1^4 2x\,dx = 15$$

Thus from (1):

$$X = \frac{93}{35} \qquad Y = \frac{45}{28}$$

and so:

$$\left(\frac{93}{35}, \frac{45}{28}\right)$$

is the required centroid.

(b)

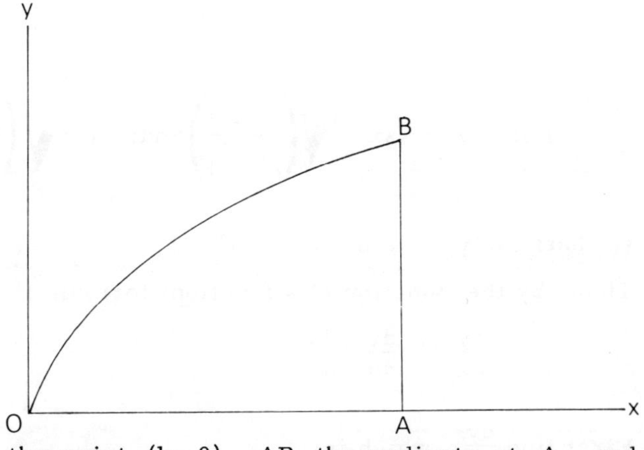

A is the point (b, 0), AB the ordinate at A, and so OAB is the area rotated about the x axis. This generates a solid of revolution whose centroid will thus lie along the x axis. Let it be the point (X, 0). The volume V of this solid is given by:

$$V = \int_0^b \pi y^2 \, . \, dx = \pi \int_0^b 4ax \, . \, dx = 2\pi ab^2$$

By the usual formula:

$$X = \frac{\int_0^b x \, . \, \pi y^2 dx}{V} = \frac{\int_0^b 4ax^2 dx}{2ab^2} = \frac{2b}{3}$$

so the centroid is the point:

$$\left(\frac{2b}{3}, 0\right)$$

15. Differentiate with respect to x:

(a) $$\sin^{-1} \sqrt{\left(\frac{x^2 - 1}{x^2 + 1}\right)}$$

(b) $$(x^2 + 1)^{\sin x}$$

(a) Let: $y = \sin^{-1} \sqrt{\left(\frac{x^2 - 1}{x^2 + 1}\right)}$ and: $u = \sqrt{\left(\frac{x^2 - 1}{x^2 + 1}\right)}$

so that: $y = \sin^{-1} u$

Then, by the 'function of a function' formula:

$$\frac{dy}{dx} = \frac{dy}{du} \cdot \frac{du}{dx} \qquad (1)$$

First find $\frac{dy}{du}$, which is:

$$\frac{1}{\sqrt{(1 - u^2)}} = \sqrt{\left(\frac{x^2 + 1}{2}\right)}$$ on reduction.

Then $\frac{du}{dx}$, which by 'quotient formula' is:

$$\frac{\sqrt{(x^2+1)} \cdot \frac{x}{\sqrt{(x^2-1)}} - \sqrt{(x^2-1)} \cdot \frac{x}{\sqrt{(x^2+1)}}}{x^2+1}$$

$$= \frac{2x}{\sqrt{(x^2-1)(x^2+1)}\,(x^2+1)}$$

Now substitution for $\frac{dy}{du}$ and $\frac{du}{dx}$ in (1) gives:

$$\frac{dy}{dx} = \frac{\sqrt{(2)} \cdot x}{(x^2+1)\sqrt{(x^2-1)}}$$

(b) Let: $y = (x^2+1)^{\sin x}$;

since: $x^2 + 1 = e^{\log(x^2+1)}$

$\therefore \quad y = e^{\sin x \cdot \log(x^2+1)}$

Let: $u = \sin x \cdot \log(x^2+1)$, so that $y = e^u$.

$$\frac{dy}{dx} = \frac{dy}{du} \cdot \frac{du}{dx} \qquad (2)$$

Now: $\frac{dy}{du} = e^u = (x^2+1)^{\sin x}$

and: $\frac{du}{dx} = \sin x \cdot \frac{2x}{x^2+1} + \cos x \cdot \log(x^2+1)$

Hence from (2):

$$\frac{dy}{dx} = (x^2 + 1)^{\sin x}\left[\sin x \cdot \frac{2x}{x^2 + 1} + \cos x \cdot \log(x^2 + 1)\right]$$

16. (a) Sketch curve $y^2 = x^2(3 - x)$, and find the area of its loop.

(b) Find the centroid of the area of the loop.

(a) First, note that the curve passes through the origin, and cuts the x axis again at the point $x = 3$ only. It cuts the y axis only at the origin. Replacing y by -y leaves the equation of the curve unaltered, so the latter is symmetrical about the x axis. Thus we need only consider positive values of y.

Since: $y = x\sqrt{(3 - x)}$

$$\frac{dy}{dx} = \frac{6 - 3x}{2\sqrt{(3 - x)}}$$

so y has a maximum when $x = 2$. Finally, since y^2 is never negative, x cannot exceed 3, and large negative values of x have large corresponding values of y^2. These facts when combined enable us to sketch the curve as follows:

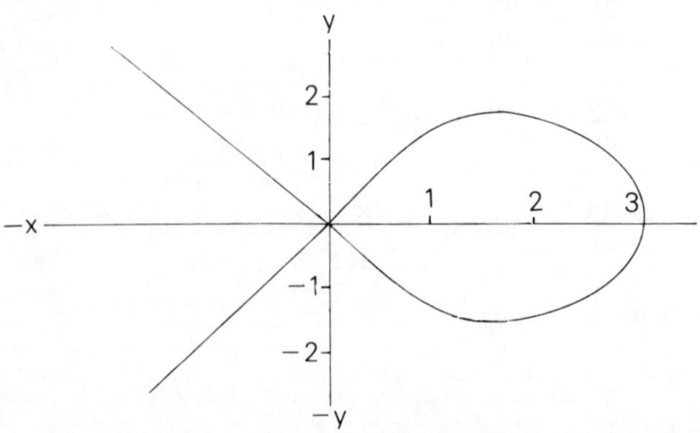

To find the area of the loop, calculate the area of it which lies above the x axis, then double this. The required area is thus:

$$2\int_0^3 y\,dx = 2\int_0^3 x \,.\, \sqrt{(3 - x)} \,.\, dx$$

To evaluate this, replace the expression under the radical, 3 - x, by u^2. This transforms the integral into:

$$4\int_0^3 (3 - u^2)u \,.\, u\,du = 4\left[u^3 - \frac{u^5}{5}\right]_0^{\sqrt{3}} = \frac{24 \,.\, \sqrt{(3)}}{5}$$

(b) By symmetry, the centroid of the loop will be on the x axis. Let it be the point $(\overline{x}, 0)$. Then by the usual formula:

$$\overline{x} = \frac{\int x\,dS}{S}$$

where dS is an element of area of the loop and S the total area. Here:

$$S = 2\int_0^3 y\,dx = 24\frac{\sqrt{(3)}}{5}$$

as has been found already. Hence:

$$\overline{x} = \frac{2\int_0^3 x^2 \,.\, \sqrt{(3 - x)} \,.\, dx}{24\frac{\sqrt{(3)}}{5}} \qquad (1)$$

To evaluate the integral:

$$\int_0^3 x^2 \,.\, \sqrt{(3 - x)} \,.\, dx$$

let: $3 - x = u^2$ and it becomes:

$$\int_0^{\sqrt{3}} (3 - u^2)^2 \,.\, u \,.\, 2u\,du$$

$$= 2\int_0^{\sqrt{3}} (9u^2 - 6u^4 + u^6)du$$

$$= 2\left[3u^3 - \frac{6u^5}{5} + \frac{u^7}{7}\right]_0^{\sqrt{3}}$$

$$= 2 . 3\sqrt{(3)}(3 - \frac{18}{5} + \frac{9}{7})$$

$$= \frac{144 . \sqrt{(3)}}{35}$$

Substitution in (1) gives:

$$\overline{x} = \frac{12}{7}$$

so the centroid of the loop is the point:

$$(\frac{12}{7}, 0)$$

17. (a) Find: $\displaystyle\int \frac{(4x + 5)dx}{\sqrt{(x^2 + 8x + 20)}}$

(b) Evaluate: $\displaystyle\int_0^{\pi/4} \frac{d\theta}{a^2 \cos^2 \theta - b^2 \sin^2 \theta}$

where a and b are constants.

(a) Let x = u + h, where h is a constant, and the integral becomes:

$$\int \frac{(4u + 4h + 5)du}{\sqrt{\{u^2 + u(2h + 8) + h^2 + 8h + 20\}}}$$

Now choose h so that the coefficient of u in the denominator, namely 2h + 8, is zero. Then h = -4, so that u = x + 4, and the last integral becomes:

$$\int \frac{(4u - 11)du}{\sqrt{(u^2 + 4)}}$$

$$= 4\int \frac{udu}{\sqrt{(u^2 + 4)}} - 11\int \frac{du}{\sqrt{(u^2 + 4)}}$$

The substitution $u^2 + 4 = v$ transforms:

$$\int \frac{udu}{\sqrt{(u^2 + 4)}} \quad \text{into:} \quad \frac{1}{2}\int v^{-\frac{1}{2}} \, . \, dv$$

$$= v^{\frac{1}{2}} = \sqrt{(u^2 + 4)}$$

also: $$\int \frac{du}{\sqrt{(u^2 + 4)}} = \log\{u + \sqrt{(u^2 + 4)}\}$$

one of the standard integrals. Hence:

$$\int \frac{(4x + 5)dx}{\sqrt{(x^2 + 8x + 20)}} = 4\sqrt{(u^2 + 4)} - 11 \, . \, \log\{u + \sqrt{(u^2 + 4)}\}$$

$$= 4\sqrt{(x^2 + 8x + 20)} -$$

$$11 \, . \, \log\{x + 4 + \sqrt{(x^2 + 8x + 20)}\}$$

on replacing u by x + 4.

(b) $$\int_0^{\pi/4} \frac{d\theta}{a^2 \cos^2 \theta - b^2 \sin^2 \theta}$$

Divide numerator and denominator by $\cos^2 \theta$, obtaining:

$$\int_0^{\pi/4} \frac{\sec^2 \theta \, . \, d\theta}{a^2 - b^2 \tan^2 \theta}$$

Let $\tan \theta = x$, so that $\sec^2 \theta d\theta$ is replaced by dx, and the integral is now:

$$\int_0^1 \frac{dx}{a^2 - b^2 x^2}$$

Since $a^2 - b^2x^2$ factorizes, this is a case for partial fractions.

As is easily found:

$$\frac{1}{a^2 - b^2 x^2} = \frac{1}{2a}\left\{\frac{1}{a - bx} + \frac{1}{a + bx}\right\}$$

so that:

$$\int_0^1 \frac{dx}{a^2 - b^2 x^2} = \frac{1}{2a}\int_0^1 \frac{dx}{a - bx} + \frac{1}{2a}\int_0^1 \frac{dx}{a + bx}$$

$$= \frac{1}{2ab}\Big[-\log(a - bx) + \log(a + bx)\Big]_0^1$$

$$= \left[\frac{1}{2ab}\log\left(\frac{a + bx}{a - bx}\right)\right]_0^1$$

Consequently:

$$\int_0^{\pi/4} \frac{d\theta}{a^2 \cos^2 \theta - b^2 \sin^2 \theta} = \frac{1}{2ab} \log\left(\frac{a + b}{a - b}\right)$$

18. (a) If: $x = a(\theta - \sin \theta)$; $y = a(1 - \cos \theta)$, prove that:

$$\frac{d^2y}{dx^2} = -\frac{a}{y^2}$$

(b) If x and y are connected by the relation $x^2 + 4xy + y^2 = 1$, find:

$$\frac{d^2y}{dx^2}$$

in terms of x and y in its simplest form.

(a) First find:

$$\frac{dy}{dx} = \frac{\frac{dy}{d\theta}}{\frac{dx}{d\theta}}$$

$$\frac{dy}{d\theta} = a \sin \theta \text{ and } \frac{dx}{d\theta} = a(1 - \cos \theta)$$

so that:

$$\frac{dy}{dx} = \frac{\sin \theta}{1 - \cos \theta}$$

Now:

$$\frac{d^2y}{dx^2} = \frac{d}{dx}\left(\frac{dy}{dx}\right) = \frac{d}{dx}\left(\frac{\sin \theta}{1 - \cos \theta}\right)$$

At this stage use the 'function of a function' formula:

$$\frac{d}{dx}\left(\frac{\sin\theta}{1 - \cos\theta}\right) = \frac{d}{d\theta}\left(\frac{\sin\theta}{1 - \cos\theta}\right) \cdot \frac{d\theta}{dx} \qquad (1)$$

$$\frac{d}{d\theta}\left(\frac{\sin\theta}{1 - \cos\theta}\right) = \frac{(1 - \cos\theta)\cos\theta - \sin^2\theta}{(1 - \cos\theta)^2}$$

$$= \frac{\cos\theta - 1}{(1 - \cos\theta)^2}$$

$$= -\frac{1}{1 - \cos\theta}$$

Also, since: $\frac{dx}{d\theta} = a(1 - \cos\theta)$

$$\frac{d\theta}{dx} = \frac{1}{a(1 - \cos\theta)}$$

And so from (1):

$$\frac{d^2y}{dx^2} = -\frac{1}{(1 - \cos\theta)} \cdot \frac{1}{a(1 - \cos\theta)}$$

$$= -\frac{1}{a(1 - \cos\theta)^2} = -\frac{a}{y^2}$$

(b) $x^2 + 4xy + y^2 = 1.$

Differentiating with respect to x:

$$2x + 4y + 4x \cdot \frac{dy}{dx} + 2y \cdot \frac{dy}{dx} = 0$$

so that: $\frac{dy}{dx} = -\frac{x + 2y}{2x + y} \qquad (2)$

Differentiating again with respect to x:

$$\frac{d^2y}{dx^2} = \frac{(x + 2y)\frac{d(2x + y)}{dx} - (2x + y)\frac{d(x + 2y)}{dx}}{(2x + y)^2}$$

$$= \frac{(x + 2y)(2 + \frac{dy}{dx}) - (2x + y)(1 + 2\frac{dy}{dx})}{(2x + y)^2}$$

i.e. $$\frac{d^2y}{dx^2} = \frac{3y - 3x\frac{dy}{dx}}{(2x + y)^2}$$

On substituting for $\frac{dy}{dx}$ from (2) this becomes:

$$\frac{3(x^2 + 4xy + y^2)}{(2x + y)^3} = \frac{3}{(2x + y)^3}$$

19. Given that:

$$u_n = \{x(1 - x)\}^n \text{ and } n \geq 2$$

prove that:

$$\frac{d^2u_n}{dx^2} = n(n - 1)u_{n-2} - 2n(2n - 1)u_{n-1}$$

Putting:

$$v_n = \int_0^1 e^x . u_n dx$$

prove that when $n \geqslant 2$:

$$v_n + 2n(2n - 1)v_{n-1} - n(n - 1)v_{n-2} = 0$$

Obtain a similar formula for:

$$w_n = \int_0^1 u_n \,.\, \cos x \, dx$$

(Ox)

$$\frac{du_n}{dx} = n\{x(1 - x)\}^{n-1} (1 - 2x)$$

$$\frac{d^2u_n}{dx^2} = n(n - 1)\{x(1 - x)\}^{n-2} \,.\, (1 - 2x)^2 - 2n\{x(1 - x)\}^{n-1}$$

by 'differentiation of product' formula. That is:

$$\frac{d^2u_n}{dx^2} = n(n - 1)\{x(1 - x)\}^{n-2} (1 - 4x + 4x^2) - 2n \,.\, u_{n-1}$$

$$= n(n - 1)u_{n-2} - n(n - 1)\{x(1 - x)\}^{n-2} \,.\, 4x(1 - x) - 2n \,.\, u_{n-1}$$

$$= n(n - 1)u_{n-2} - 4n(n - 1)\{x(1 - x)\}^{n-1} - 2n \,.\, u_{n-1}$$

$$= n(n - 1)u_{n-2} - 4n(n - 1)u_{n-1} - 2n \,.\, u_{n-1}$$

$$= n(n-1)u_{n-2} - 2n(2n-1)u_{n-1}$$

Now for:

$$v_n = \int_0^1 e^x . u_n \, dx = \int_0^1 u_n . \frac{de^x}{dx} . dx$$

Integrate by parts and get:

$$v_n = \left[e^x u_n \right]_0^1 - \int_0^1 e^x \frac{du_n}{dx} . dx$$

u_n vanishes when x = 0 and x = 1, so that:

$$v_n = - \int_0^1 e^x \frac{du_n}{dx} . dx$$

$$= - \left[e^x \frac{du_n}{dx} \right]_0^1 + \int_0^1 e^x \frac{d^2u_n}{dx^2} . dx$$

Now $\frac{du_n}{dx}$, also, vanishes when x = 0 and x = 1. Thus:

$$v_n = \int_0^1 e^x \frac{d^2u_n}{dx^2} . dx$$

$$= \int_0^1 e^x \left[n(n-1)u_{n-2} - 2n(2n-1)u_{n-1} \right] dx$$

$$= n(n-1)v_{n-2} - 2n(2n-1)v_{n-1}$$

i.e. $v_n + 2n(2n-1)v_{n-1} - n(n-1)v_{n-2} = 0.$

Finally:

$$w_n = \int_0^1 u_n \cdot \cos x \, dx$$

Integrating by parts:

$$w_n = \left[u_n \sin x\right]_0^1 - \int_0^1 \sin x \cdot \frac{du_n}{dx} \cdot dx$$

$$= -\int_0^1 \sin x \cdot \frac{du_n}{dx} \cdot dx$$

Integrating again by parts, this last integral gives:

$$w_n = \left[\cos x \cdot \frac{du_n}{dx}\right]_0^1 - \int_0^1 \cos x \cdot \frac{d^2u_n}{dx^2} \cdot dx$$

$$= -\int_0^1 \cos x \cdot \frac{d^2u_n}{dx^2} \cdot dx$$

$$= -\int_0^1 \cos x \{n(n-1)u_{n-2} - 2n(2n-1)\, u_{n-1}\} dx$$

$$= -n(n-1)w_{n-2} + 2n(2n-1)w_{n-1}$$

That is:

$$w_n - 2n(2n-1)w_{n-1} + n(n-1)w_{n-2} = 0$$

20. From a given amount of thin metal of area S cm^2 it is desired to construct a container in the shape of a right circular cylinder (with lid) having the maximum volume. Find, in terms of S what the height and radius of the container must be, and the maximum volume.

Consider a container of this shape whose radius is r cm and height h cm. Its total surface area is $2\pi r^2 + 2\pi rh$. And so under the given conditions:

$$2\pi r^2 + 2\pi rh = S \qquad (1)$$

If V cm^3 is the volume of the container:

$$V = \pi r^2 . h \qquad (2)$$

which is to be a maximum. Though V contains two variables, r and h, we can eliminate h. From (1):

$$\pi rh = \frac{S - 2\pi r^2}{2}$$

Substituting in (2) expresses V as a function of one variable only, r.

$$V = r . \pi rh$$

$$= r\frac{(S - 2\pi r^2)}{2}$$

$$\therefore \quad \frac{dV}{dr} = \frac{1}{2}(S - 2\pi r^2 - 4\pi r^2) = 0 \quad \text{when } 6\pi r^2 = S$$

i.e. when:

$$r = \left(\frac{S}{6\pi}\right)^{\frac{1}{2}} \text{ cm.} \qquad \text{Since: } \frac{d^2V}{dr^2}$$

is negative, this value of r makes V a maximum. To find the corresponding value of h, from (1):

$$h = \frac{S - 2\pi r^2}{2\pi r} \quad \text{and:} \quad S = 6\pi r^2$$

$$\text{so:} \quad h = \frac{4\pi r^2}{2\pi r} = 2r = 2\left(\frac{S}{6\pi}\right)^{\frac{1}{2}} \text{ cm}$$

From (2), the maximum volume of the container will be:

$$\pi r^2 . 2r = 2\pi r^2 . r = \frac{S}{3}\left(\frac{S}{6\pi}\right)^{\frac{1}{2}} = \frac{S^{3/2}}{3 . (6\pi)^{\frac{1}{2}}} \text{ cm}^3$$

21. Given that u, v are two distinct functions of x, such that uv = a, where a is a non zero constant, and that:

(i) $$\frac{d^2u}{dx^2} + p(x)\frac{du}{dx} + q(x)u = 0$$

(ii) $$\frac{d^2v}{dx^2} + p(x)\frac{dv}{dx} + q(x)v = 0$$

prove that:

(a) $$u\frac{d^2v}{dx^2} + 2\frac{du}{dx} . \frac{dv}{dx} + v\frac{d^2u}{dx^2} = 0$$

(b) $$\frac{du}{dx} . \frac{dv}{dx} = aq(x)$$

(c) $$\frac{dq(x)}{dx} + 2p(x)q(x) = 0$$

(Ox)

For short denote:

$\frac{du}{dx}, \frac{dv}{dx}$ by u' and v', $\frac{d^2u}{dx^2}, \frac{d^2v}{dx^2}$ by u'' and v''

(a) Since $uv = a$, differentiating we get:

$$uv' + u'v = 0$$

and differentiating this:

$$uv'' + u'v' + u'v' + u''v = 0$$

i.e. $$uv'' + 2u'v' + u''v = 0$$

so (a) is proved.

(b) Multiply (i) by v, (ii) by u and add. We get:

$$v(u'' + p(x)u' + q(x)u) + u(v'' + p(x)v' + q(x)v) = 0 \qquad (1)$$

But from (a):

$$vu'' + uv'' = -2u'v'$$

also, as was shown:

$$uv' + u'v = 0$$

Thus (1) becomes after simplification:

$$-2u'v' + 2q(x)uv = 0$$ and so:

$$u'v' = q(x)uv = a \, . \, q(x)$$

Thus (b) is proved.

(c) Differentiating (b) gives:

$$a\frac{dq(x)}{dx} = u'v'' + u''v'$$

Multiply (i) by v', (ii) by u' and add. This gives:

$$u'(v'' + p(x)v' + q(x)v) + v'(u'' + p(x)u' + q(x)u) = 0$$

Multiply out, noting that $u'v + uv' = 0$. The result is:

$$u'v'' + v'u'' = -2p(x)u'v'$$

so that:

$$a\frac{dq(x)}{dx} = u'v'' + v'u'' = -2p(x)u'v'$$

$$= -2p(x) \,.\, a \,.\, q(x) \text{ from (b).}$$

Therefore dividing across by a, since $a \neq 0$, we get:

$$\frac{dq(x)}{dx} = -2p(x) \,.\, q(x)$$

or: $$\frac{dq(x)}{dx} + 2p(x) \,.\, q(x) = 0$$

22. PQ is a variable chord of the ellipse:

$$\frac{x^2}{a^2} + \frac{y^2}{b^2} = 1$$

and O is the origin. If the angle POQ is always a right angle, and ON is drawn perpendicular to PQ, find the locus of N.

Let the equation of PQ, in one of its positions, be $x \cos \alpha + y \sin \alpha = p$, so that ON = p. Consider the co-ordinates of P. They satisfy:

$$\frac{x^2}{a^2} + \frac{y^2}{b^2} = 1$$

because P is on the ellipse, and also satisfy:

$$\frac{x \cos \alpha + y \sin \alpha}{p} = 1$$

because P is on the line $x \cos \alpha + y \sin \alpha = p$.

Thus the co-ordinates of P, and similarly those of Q, satisfy this homogeneous equation of second degree in x and y-:

$$\frac{x^2}{a^2} + \frac{y^2}{b^2} = \left(\frac{x \cos \alpha + y \sin \alpha}{p}\right)^2 \qquad \text{or:}$$

$$\left(\frac{1}{a^2} - \frac{\cos^2 \alpha}{p^2}\right)x^2 - \frac{2xy \cos \alpha \sin \alpha}{p^2} + \left(\frac{1}{b^2} - \frac{\sin^2 \alpha}{p^2}\right)y^2 = 0 \qquad (1)$$

This has factors, of form:

$$\left(\frac{1}{b^2} - \frac{\sin^2 \alpha}{p^2}\right)(y - m_1 x)(y - m_2 x) = 0 \qquad (2)$$

which represents a pair of straight lines through the origin, these being OP and OQ. Now OP and OQ are at right angles, so that $m_1 \cdot m_2 = -1$. Hence, equating the coefficients of x^2 in (1) and (2) we get:

$$\frac{1}{a^2} - \frac{\cos^2 \alpha}{p^2} = \left(\frac{1}{b^2} - \frac{\sin^2 \alpha}{p^2}\right)m_1 m_2 = -\left(\frac{1}{b^2} - \frac{\sin^2 \alpha}{p^2}\right)$$

i.e: $$\frac{1}{a^2} + \frac{1}{b^2} = \frac{\cos^2 \alpha + \sin^2 \alpha}{p^2} = \frac{1}{p^2}$$

Therefore: $$p^2 = \frac{a^2 b^2}{a^2 + b^2}$$

so that the locus of N is the circle:

$$x^2 + y^2 = \frac{a^2 b^2}{a^2 + b^2}$$

23. The parabola $y^2 = 4ax$ and the hyperbola $4xy = a^2$ meet at P. Their common tangent touches the parabola at Q and the hyperbola at R. Prove that PR is the tangent at P to the parabola, and that QP is the tangent at P to the hyperbola.

(Ox)

Any point on the parabola has co-ordinates $(at^2, 2at)$, and any point on the hyperbola has co-ordinates:

$$\left(a\frac{T}{2}, \frac{a}{2T}\right)$$

It is easily seen that P is the point: $\left(\frac{a}{4}, a\right)$.

Let the common tangent to the parabola and hyperbola touch them at the points 't' and 'T' respectively.

Then, from the parabola, the equation of the common tangent will be:

$$y - \frac{1}{t} \cdot x - at = 0$$

But, from the hyperbola, the equation will also be:

$$y + \frac{x}{T^2} - \frac{a}{T} = 0$$

As these are identical:

$$-\frac{1}{t} = \frac{1}{T^2} \qquad \text{and: } t = \frac{1}{T}$$

And so $T^2 = -\frac{1}{T}$, i.e. $T = -1$, so $t = -1$.

Hence Q is the point $(a, -2a)$, and R the point $\left(-\frac{a}{2}, -\frac{a}{2}\right)$.

Thus the equation of PR is:

$$\frac{y - a}{a + \frac{a}{2}} = \frac{x - \frac{a}{4}}{\frac{a}{4} + \frac{a}{2}}$$

which reduces to:

$$y = 2x + \frac{a}{2}$$

the equation of the tangent at P to the parabola.

Now for the equation of QP. This is:

$$\frac{x - a}{a - \frac{a}{4}} = \frac{y + 2a}{-2a - a}$$

which reduces to:

$$y + 4x - 2a = 0$$

the equation of the tangent to the hyperbola at P.

24. A triangular pyramid has an equilateral triangle ABC of side a as base and the lines joining A, B, C to the vertex V are all of length b.

(a) Find the length of the perpendicular from B to VA, and hence or otherwise find the angle between the planes VAB and VAC.

(b) Find the length of the shortest distance between BC and VA.

(Ox)

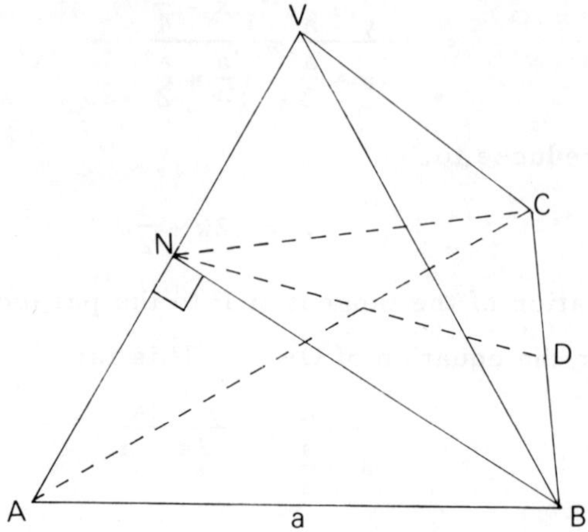

(a) Draw BN perpendicular to VA, and let BN = p, also let $\angle VAB = \alpha$. Then $BN = AB \sin\alpha = a \sin\alpha$. Now from isosceles triangle VAB:

$$\cos\alpha = \frac{AB}{2AV} = \frac{a}{2b}$$

Hence: $$p = a\sin\alpha = a\sqrt{(1 - \cos^2\alpha)}$$

$$= a\sqrt{\left(1 - \frac{a^2}{4b^2}\right)} = \frac{a}{2b}\sqrt{(4b^2 - a^2)}$$

From symmetry CN will be the perpendicular to VA from C and CN = p, also.

Consider the planes VAB, VAC. They intersect in the line VA, N is a point on VA and NB, NC are drawn perpendicular to this line, one in each plane. Therefor $\angle BNC$ is the angle between these two planes, 2β say. Join N to D, the middle point of BC. Since triangle BNC is isosceles, ND is at right angles to

BC and $\angle$BND = $\angle$DNC = β. Also:

$$\sin\beta = \frac{BD}{BN} = \frac{a}{2} \div \frac{a}{2b}\sqrt{(4b^2 - a^2)}$$

$$= \frac{b}{\sqrt{(4b^2 - a^2)}}$$

$\therefore$ the angle between planes VAB, VAC = 2β where:

$$\sin\beta = \frac{b}{\sqrt{(4b^2 - a^2)}}$$

(b) Since VA is perpendicular to both NB and NC, it will be perpendicular to every line in their plane, in particular to ND. Consequently ND, being perpendicular to and intersecting both BC and VA, will be the shortest distance between these lines.

From right-angled triangle BND:

$$ND^2 = BN^2 - BD^2$$

$$= \frac{a^2}{4b^2}(4b^2 - a^2) - \frac{a^2}{4} = \frac{3a^2}{4} - \frac{a^4}{4b^2}$$

$$\therefore \quad ND = \frac{a}{2b}\sqrt{(3b^2 - a^2)}$$

We may also find ND from the fact that ND = NB cos β, and sin β is known.

25. A and B are two given points $(x_1 y_1)$ and $x_2 y_2$ respectively. Through A a line APQ is drawn to meet the ellipse:

$$\frac{x^2}{a^2} + \frac{y^2}{b^2} = 1$$

in P and Q.

Through B a line BRS is drawn parallel to APQ to meet the ellipse in R and S. Prove that the ratio of the rectangles AP . AQ and BR . BS is constant, independent of the directions of APQ and BRS.

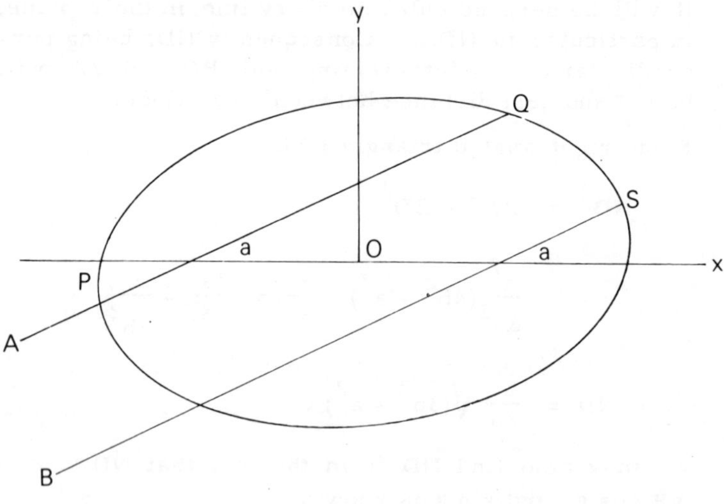

Let O be the origin, and suppose that APQ and BRS each make an angle α with the positive x axis. Then the co-ordinates of any point N on APQ will be $x_1 + r \cos \alpha$, $y_1 + r \sin \alpha$, where r is the distance AN. To find the distances AP and AQ substitute $x_1 + r \cos \alpha$ for x, $y_1 + r \sin \alpha$ for y in the equation of the ellipse, obtaining:

$$\frac{(x_1 + r\cos\alpha)^2}{a^2} + \frac{(y_1 + r\sin\alpha)^2}{b^2} = 1$$

This is a quadratic equation for r, namely:

$$r^2(b^2\cos^2\alpha + a^2\sin^2\alpha) + 2r(b^2x_1\cos\alpha + a^2y_1\sin\alpha)$$

$$+ b^2x_1^2 + a^2y_1^2 - a^2b^2 = 0$$

The two roots of this will be the lengths AP and AQ, and so:

$$AP \cdot AQ = \frac{b^2x_1^2 + a^2y_1^2 - a^2b^2}{b^2\cos^2\alpha + a^2\sin^2\alpha} \qquad (1)$$

Similarly:

$$BR \cdot BS = \frac{b^2x_2^2 + a^2y_2^2 - a^2b^2}{b^2\cos^2\alpha + a^2\sin^2\alpha} \qquad (2)$$

Division of (1) by (2) gives:

$$\frac{AP \cdot AQ}{BR \cdot BS} = \frac{b^2x_1^2 + a^2y_1^2 - a^2b^2}{b^2x_2^2 + a^2y_2^2 - a^2b^2}$$

which is independent of α, i.e. of the directions of APQ and BRS.

26. A, B are two points on the rectangular hyperbola $xy = c^2$. The circle on AB as diameter cuts the hyperbola again at C, D. Prove that CD goes through the origin. AB meets CD at P. Prove that if AB varies but passes always through the fixed point (h, k) then P lies on the hyperbola:

$$x(x - h) - y(y - k) = 0$$

(Ox)

Let A, B be the points:

$$\left(ct_1, \frac{c}{t_1}\right)\left(ct_2, \frac{c}{t_2}\right)$$

on the hyperbola. Suppose C is the point:

$$\left(ct, \frac{c}{t}\right)$$

Then, from the given conditions, CA and CB are at right angles, so that:

$$\frac{\frac{c}{t} - \frac{c}{t_1}}{ct - ct_1} \cdot \frac{\frac{c}{t} - \frac{c}{t_2}}{ct - ct_2} = -1$$

which after simplifying becomes:

$$t^2 = -\frac{1}{t_1 t_2} \qquad (1)$$

Let the two values of t obtained from this be T_1 and T_2, so that C will be the point:

$$\left(cT_1, \frac{c}{T_1}\right)$$

D the point:

$$\left(cT_2, \frac{c}{T_2}\right)$$

The line CD will pass through the origin O if OC and OD have the same gradient. This will be true if:

$$\frac{\frac{c}{T_1}}{cT_1} = \frac{\frac{c}{T_2}}{cT_2}$$

i.e. if: $$T_1^2 = T_2^2$$

the truth of which follows from (1).

The equation of CD will therefore be:

$$y = \frac{x}{T_1^2} = -t_1 t_2 x \qquad (2)$$

The equation of AB is:

$$\frac{y - \frac{c}{t_1}}{\frac{c}{t_1} - \frac{c}{t_2}} = \frac{x - ct_1}{ct_1 - ct_2}$$

which reduces to:

$$yt_1 t_2 + x - c(t_1 + t_2) = 0 \qquad (3)$$

Since AB passes through the fixed point (h, k):

$$kt_1 t_2 + h - c(t_1 + t_2) = 0 \qquad (4)$$

Solving for x and y between (2) and (3) we get X and Y, the co-ordinates of P, the point where AB meets CD.

They are found to be:

$$X = \frac{c(t_1 + t_2)}{1 - t_1^2 t_2^2} \qquad (5)$$

$$Y = -t_1 t_2 . X \qquad (6)$$

The locus of P is obtained by eliminating $t_1 t_2$ and $t_1 + t_2$ from (4), (5) and (6).

From (5):

$$c(t_1 + t_2) = X(1 - t_1^2 t_2^2)$$

$$= X(1 - \frac{Y^2}{X^2}) \text{ from (6)}$$

$$= X - \frac{Y^2}{X}$$

Thus, substituting in (4):

$$-k . \frac{Y}{X} + h - (X - \frac{Y^2}{X}) = 0$$

i.e. $$X^2 - Y^2 - Xh + Yk = 0$$

and so the locus of P is the hyperbola:

$$x(x - h) - y(y - k) = 0$$

27. Spheres are drawn each passing through two fixed points P and Q, and touching a fixed plane p. P and Q are on the same side of p. Prove that the locus of the point of contact of the spheres with the plane is a circle whose centre is the point of intersection of PQ with p.

Prove also that the centres of the spheres lie on the curve of intersection of a certain right circular cylinder, whose axis is perpendicular to p, with a certain plane perpendicular to PQ.

(Ox)

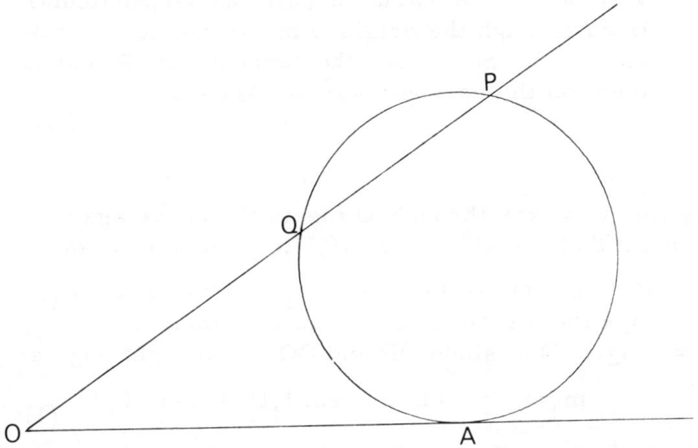

Let PQ meet the plane p at O, and consider any one sphere through P and Q and touching p at A. Then the plane POA cuts this sphere in a circle, to which OA is a tangent and OQP a secant. From the geometry of the circle:

$$OA^2 = OP \cdot OQ$$

Since OP and OQ are given lengths, it follows that OA is of known length, so that the locus of A is a circle whose centre is O.

The centre of the sphere must lie on (1) the plane which bisects the chord PQ at right angles; (2) the line drawn perpendicular to the plane p at A. Now as A lies on a given circle, the line perpendicular to p at A will trace out the surface of a right circular cylinder whose axis is perpendicular to p.

Therefore the centres of the spheres lie on the curve in which this right circular cylinder intersects the plane bisecting PQ at right angles.

28. A curve is given parametrically by $x = at^2$, $y = at^3$. A variable pair of perpendicular lines through the origin O meets the curve at P and Q. Show that the tangents at P and Q meet on the parabola $4y^2 = 3ax - a^2$.

(Ox)

Any line $y = mx$ through O meets the curve again at the point 't' where $at^3 = m\,at^2$, i.e. where $t = m$.

Thus if P and Q are the points 't_1' and 't_2', $y = m_1 x$, $y = m_2 x$ the equations of OP and OQ, then $t_1 = m_1$, $t_2 = m_2$. Also since OP and OQ are at right angles:

$$m_1 m_2 = -1 \qquad \text{so: } t_1 t_2 = -1 \qquad (1)$$

The equation of the tangent to the curve at P is:

$$\frac{y - at_1^3}{x - at_1^2} = \frac{3at_1^2}{2at_1} = \frac{3t_1}{2}$$

or: $$2y - 3t_1 x + at_1^3 = 0 \qquad (2)$$

and the equation of the tangent at Q is:

$$2y - 3t_2 x + at_2^3 = 0 \qquad (3)$$

The point (X, Y) where these tangents meet is found by solving for x and y from (2) and (3).

Hence: $$X = \frac{a}{3}(t_1^2 + t_2^2 + t_1 t_2)$$

$$= \frac{a}{3}(t_1^2 + t_2^2 - 1) \text{ from (1)} \qquad (4)$$

and: $$Y = -\frac{a}{2}(t_1 + t_2) \qquad (5)$$

To find the locus of (X, Y) eliminate t_1 and t_2 between (1), (4) and (5). Squaring (5):

$$\frac{4Y^2}{a^2} = t_1^2 + t_2^2 + 2t_1t_2 = t_1^2 + t_2^2 - 2$$

From (4):

$$\frac{3X}{a} = t_1^2 + t_2^2 - 1$$

Therefore:

$$\frac{3X}{a} - 1 = \frac{4Y^2}{a^2}$$

Replacing X, Y by 'running' co-ordinates (x, y), the equation of the required locus of (X, Y) is seen to be the parabola:

$$4y^2 = 3ax - a^2$$

29. Show that the ellipse:

$$\frac{x^2}{p^2} + \frac{y^2}{q^2} = 2$$

touches the hyperbola $xy = pq$ at the points (p, q), (-p, -q). If the tangent to the ellipse at a point P on it meets the hyperbola at R and S, prove that the tangents to the hyperbola at R and S meet at a point Q on the ellipse.

Prove also that if the tangent at Q to the ellipse meets the hyperbola at U and V, then PU and PV are the tangents to the hyperbola at U and V.

(Ox)

Solving for x between the equations:

$$\frac{x^2}{p^2} + \frac{y^2}{q^2} = 2 \text{ and } xy = pq$$

we get:

$$\frac{x^2}{p^2} + \frac{p^2}{x^2} = 2$$

so $x = p, -p$, $y = q, -q$, i.e. the curves touch at (p, q), $(-p, -q)$.

Any point on the hyperbola can be expressed in form:

$$x = pt, \; y = \frac{q}{t}$$

Take the point P on the ellipse as:

$$(\sqrt{(2)} \, . \, p \cos a, \; \sqrt{(2)} \, . \, q \sin a)$$

so that the equation of the tangent at P will be:

$$\frac{x\sqrt{2} \cos a}{p} + \frac{y\sqrt{2} \sin a}{q} = 2$$

or: $$\frac{x \cos a}{p} + \frac{y \sin a}{q} = \sqrt{(2)}$$

This will meet the hyperbola at the points:

$$R(pt_1, \frac{q}{t_1}) \text{ and } S(pt_2, \frac{q}{t_2})$$

where t_1 and t_2 are the roots of the equation:

$$\frac{pt \cos a}{p} + \frac{q \sin a}{tq} = \sqrt{(2)}$$

or: $t^2 \cos a - \sqrt{(2)} \, . \, t + \sin a = 0$ (1)

The equations of the tangents to the hyperbola at R and S are found to be:

$$x\frac{q}{t_1} + y \cdot pt_1 = 2pq$$

and: $$x\frac{q}{t_2} + y \cdot pt_2 = 2pq$$

Solving these to find their point of intersection Q(X, Y) we get:

$$X = \frac{2pt_1t_2}{t_1 + t_2} \qquad Y = \frac{2q}{t_1 + t_2} \qquad (2)$$

To show that Q lies on the ellipse: from (1):

$$t_1 + t_2 = \frac{\sqrt{(2)}}{\cos a} \qquad t_1t_2 = \frac{\sin a}{\cos a}$$

so from (2):

$$X = \sqrt{(2)}\, p \sin a,\ Y = \sqrt{(2)}\, q \cos a$$

$$\frac{X^2}{p^2} + \frac{Y^2}{q^2} = 2$$

i.e. Q lies on the ellipse.

The tangent to the ellipse at Q is:

$$x\frac{\sqrt{(2)}\sin a}{p} + y\frac{\sqrt{(2)}\cos a}{q} = 2$$

or: $$\frac{x \sin a}{p} + \frac{y \cos a}{q} = \sqrt{(2)} \qquad (3)$$

To find where this meets the hyperbola, put:

$$x = pu \qquad y = \frac{q}{u}$$

in (3), obtaining the quadratic equation:

$$u^2 \sin a - \sqrt{(2)}\, u + \cos a = 0 \qquad (4)$$

with roots u_1 and u_2.

And so the tangent at Q to the ellipse meets the hyperbola in two points:

$$U\left(pu_1, \frac{q}{u_1}\right), \quad V\left(pu_2, \frac{q}{u_2}\right)$$

where, from (4):

$$u_1 + u_2 = \frac{\sqrt{(2)}}{\sin a} \qquad u_1 u_2 = \frac{\cos a}{\sin a} \tag{5}$$

Now, as before, the tangents to the hyperbola at the points U and V intersect at the point:

$$\left(\frac{2pu_1u_2}{u_1 + u_2}, \frac{2q}{u_1 + u_2}\right)$$

which, from (5), is the point:

$$(\sqrt{(2)}\ p \cos a, \sqrt{(2)}\ q \sin a)$$

i.e. P.

Therefore PU and PV are the tangents to the hyperbola at U and V.

30. A variable chord AB of the parabola $y^2 = 4ax$ is of constant length l. Find the locus of its middle point.

The co-ordinates of any point on this parabola can be stated in terms of a parameter t by the usual equations:

$$x = at^2 \qquad y = 2at$$

Let A and B be the points:

$$(at_1^2, 2at_1), (at_2^2, 2at_2)$$

Then:

$$AB^2 = (at_1^2 - at_2^2)^2 + (2at_1 - 2at_2)^2$$

$$= a^2(t_1 - t_2)^2(t_1 + t_2)^2 + 4a^2(t_1 - t_2)^2 = 1^2$$

by the given condition. This may be written as:

$$(t_1 - t_2)^2\left\{(t_1 + t_2)^2 + 4\right\} = \frac{1^2}{a^2} \tag{1}$$

Let (X, Y) be the middle point of AB. Then:

$$X = \frac{a(t_1^{\ 2} + t_2^{\ 2})}{2} \tag{2}$$

and: $$Y = \frac{2a(t_1 + t_2)}{2} = a(t_1 + t_2) \tag{3}$$

The locus of (X, Y) is got by eliminating t_1 and t_2 between equations (1) to (3).

The identity:

$$(t_1 - t_2)^2 + (t_1 + t_2)^2 \equiv 2(t_1^{\ 2} + t_2^{\ 2})$$

enables us to find $(t_1 - t_2)^2$ in terms of X and Y. From (2):

$$t_1^{\ 2} + t_2^{\ 2} = \frac{2X}{a}$$

From (3):

$$(t_1 + t_2)^2 = \frac{Y^2}{a^2}$$

Thus:

$$(t_1 - t_2)^2 + \frac{Y^2}{a^2} = \frac{4X}{a}$$

or: $$(t_1 - t_2)^2 = \frac{4aX - Y^2}{a^2}$$

Accordingly, substituting in (1) it becomes:

$$\frac{(4aX - Y^2)}{a^2}\left(\frac{Y^2}{a^2} + 4\right) = \frac{l^2}{a^2}$$

or: $$(Y^2 - 4aX)(Y^2 + 4a^2) + a^2l^2 = 0$$

Replacing X, Y by 'running' co-ordinates x, y gives the required locus of the middle point of AB, namely:

$$(y^2 - 4ax)(y^2 + 4a^2) + a^2l^2 = 0$$

31. Solve the equations:

(a) $x + y = 6.$

(b) $xz + yw = -2.$

(c) $xz^2 + yw^2 = 22.$

(d) $xz^3 + yw^3 = -50.$

for x, y, z, w.

Multiply (a) by (c) and subtract the square of (b):

$$\therefore \qquad xy(z - w)^2 = 128 \qquad (1)$$

Multiply (b) by (d) and subtract the square of (c):

then: $$xyzw(z - w)^2 = -384 \qquad (2)$$

Division of (2) by (1) gives:

$$zw = -3 \qquad (3)$$

Multiply (a) by (d), then (b) by (c) and subtract the two products. As a result:

$$xy(z^3 + w^3 - z^2w - zw^2) = -256$$

i.e. $$xy(z + w)(z - w)^2 = -256 \qquad (4)$$

Division of (4) by (1) gives:

$$z + w = -2 \qquad (5)$$

Therefore, from (3) and (5), either:

$$z = 1, w = -3; \quad \text{or:} \quad z = -3, w = 1$$

Taking the first pair of values, $z = 1$, $w = -3$, and then solving (a) and (b) for x and y, we find that $x = 4$, $y = 2$. If instead $z = -3$, $w = 1$, then $x = 2$, $y = 4$.

32. Each of the numbers $x_1, x_2 \ldots x_k$ has one or other of two non zero values:

$$\alpha, \beta, \text{ and } \sum_{n=1}^{k} x_n = 0$$

$$\sum_{n=1}^{k} x_n^2 = 1$$

Prove that if V is written for:

$$\sum_{n=1}^{k} x_n^3$$

then V has one of the 2k - 2 values:

$$\pm \frac{1}{\sqrt{(k)}} \left\{ \sqrt{\left(\frac{k - r}{r}\right)} - \sqrt{\left(\frac{r}{k - r}\right)} \right\}$$

where r = 1, 2 . . . k - 1.

Hence show that the greatest value of V is:

$$\frac{k-2}{\sqrt{(k(k-1))}}$$

(Ox)

Suppose that, of the numbers $x_1, x_2 \ldots x_k$, r are alphas and k - r are betas. Then by hypothesis:

$$r\alpha + (k-r)\beta = 0 \qquad (1)$$

$$r\alpha^2 + (k-r)\beta^2 = 1 \qquad (2)$$

and:

$$V = r\alpha^3 + (k-r)\beta^3 \qquad (3)$$

From (1) we can let:

$$\alpha = \frac{\lambda}{r} \qquad \beta = -\frac{\lambda}{k-r}$$

Substituting in (2) we get:

$$\frac{\lambda^2}{r} + \frac{\lambda^2}{k-r} = 1$$

so that:

$$\lambda = \pm\sqrt{\left(\frac{r(k-r)}{k}\right)}$$

This gives:

$$\alpha = \frac{\lambda}{r} = \pm\sqrt{\left(\frac{k-r}{kr}\right)}$$

$$\beta = \pm\sqrt{\left(\frac{r}{k(k-r)}\right)}$$

Substituting for α and β in (3), it becomes:

$$V = \pm\left(r\left(\frac{k-r}{kr}\right)\sqrt{\left(\frac{k-r}{kr}\right)} - \frac{(k-r)r}{k(k-r)}\sqrt{\left(\frac{r}{k(k-r)}\right)}\right)$$

$$= \pm \frac{r(k-r)}{k}\left(\frac{1}{r}\sqrt{\left(\frac{k-r}{kr}\right)} - \frac{1}{k-r}\sqrt{\left(\frac{r}{k(k-r)}\right)}\right)$$

$$= \pm \frac{r(k-r)}{k}\left(\frac{(k-r)^2 - r^2}{r(k-r)\sqrt{(kr(k-r))}}\right)$$

$$= \pm \frac{k-2r}{\sqrt{(kr(k-r))}}$$

$$= \pm \frac{(k-r)-r}{\sqrt{(kr(k-r))}}$$

$$= \pm \frac{1}{\sqrt{(k)}}\left(\sqrt{\left(\frac{k-r}{r}\right)} - \sqrt{\left(\frac{r}{k-r}\right)}\right) \qquad (4)$$

As neither α nor β is zero, it follows from (1) that, here, r cannot have the values 0 or k, hence r = 1, 2 . . . k - 1. Thus (4) provides 2k - 2 values for V. To find the greatest value of V, let:

$$\sqrt{\left(\frac{k-r}{r}\right)} = x$$

then: $\sqrt{\left(\frac{k-r}{r}\right)} - \sqrt{\left(\frac{r}{k-r}\right)} = x - \frac{1}{x}$

The differential coefficient of this with respect to x is:

$$1 + \frac{1}{x^2}$$

which is always positive so:

$$x - \frac{1}{x}$$

is steadily increasing. Its maximum value is, therefor when x, i.e:

$$\frac{k-r}{r}$$

is a maximum, which is when r = 1. Therefore the maximum value of V is:

$$\frac{1}{\sqrt{(k)}}\left(\sqrt{\left(\frac{k-1}{1}\right)} - \sqrt{\left(\frac{1}{k-1}\right)}\right) = \frac{k-2}{\sqrt{(k(k-1))}}$$

33. Find what ranges of values of x satisfy the inequality:

$$\frac{(2x-3)(2x-1)}{x(x-4)} \geqslant 3$$

$$\frac{(2x-3)(2x-1)}{x(x-4)} - 3 \geqslant 0$$

On reducing:

$$\frac{x^2 + 4x + 3}{x(x-4)}$$

i.e. $$\frac{(x+3)(x+1)}{x(x-4)} \geqslant 0 \qquad (1)$$

Equality in (1) takes place only when $x = -3$ or $x = -1$. For other values of x, (1) is true if, and only if, both numerator and denominator have the same sign.

First, suppose that this sign is positive, so that:

$$(x+3)(x+1) > 0 \qquad (2)$$

$$x(x-4) > 0 \qquad (3)$$

Now (2) is satisfied when:

$$x > -1 \text{ or } x < -3$$

and (3) is satisfied when:

$$x > 4 \text{ or } x < 0$$

Thus (2) and (3), and consequently (1), are both satisfied when $x > 4$, or $x < -3$, or $-1 < x < 0$.

Next suppose that, instead of (2) and (3), we have:

$$(x + 3)(x + 1) < 0 \qquad (4)$$

$$x(x - 4) < 0 \qquad (5)$$

Then (1) is still true.

(4) is satisfied if: $-3 < x < -1$.

(5) is satisfied if: $0 < x < 4$.

Clearly (4) and (5) cannot both be satisfied for the same values of x.

Therefore the ranges of values of x which satisfy the given inequality are:

$$x > 4; \quad -1 \leq x < 0; \quad x \leq -3$$

34. A bag contains 3n balls. Of these, n are all white and indistinguishable from one another, a further n are all black and indistinguishable from one another, and the remaining n all have colour markings different from one another and from the black and white balls. Prove that the number of distinguishable selections of n balls that can be made, all the balls being replaced after each selection is:

$$^{n}C_{n} + 2 \,.\, ^{n}C_{n-1} + 3 \,.\, ^{n}C_{n-2} + \ldots + (n+1) \,.\, ^{n}C_{0}$$

where $^{n}C_{r}$ is the number of combinations of n things taken r at a time. Prove that this expression is equal to:

$$2^{n-1} \,.\, (n + 2)$$

(Ox)

For short refer to the three sets of n balls as (1)n white balls, (2)n black balls, (3)n with colour markings different from one another and from the black and white balls.

The number of distinguishable selections of n balls from (3) is 1, or $^{n}C_{n}$. The number of selections of (n - 1) balls from (3) is $^{n}C_{n-1}$. Since the remaining ball can be selected in two ways (either from (1) or (2)), the total number of selections of n balls, with (n - 1) being from (3) is $2 \cdot {}^nC_{n-1}$.

Generally the number of selections of (n - r) balls from (3) is ${}^nC_{n-r}$. The remaining r balls can then be selected in (r + 1) ways (according as 0, 1, 2, 3 . . r of them are selected from (1) and the remainder from (2)). And so the number of selections of n balls of which n - r are selected from (3) and the remaining r balls from (1) and (2) is $(r + 1) . {}^nC_{n-r}$, where r has the values 0, 1, 2, 3 . . . n, in turn.

Adding up, the required number of distinguishable selections of n balls is:

$$ {}^nC_n + 2 . {}^nC_{n-1} + 3 . {}^nC_{n-2} + . . . + (n+1) . {}^nC_0 $$

To show this is equal to $2^{n-1}(n + 2)$ a device is used. From the binomial expansion:

$$ {}^nC_n + {}^nC_{n-1} . x + {}^nC_{n-2} . x^2 + . . . + {}^nC_{n-r} . x^r $$

$$ + . . . + {}^nC_0 . x^n \equiv (1 + x)^n $$

Multiply both sides by x:

$$ {}^nC_n . x + {}^nC_{n-1} . x^2 + {}^nC_{n-2} . x^3 + . . . + {}^nC_{n-r} . x^{r+1} $$

$$ + . . . + {}^nC_0 . x^{n+1} \equiv x(1 + x)^n $$

Differentiate both sides with respect to x:

$$ {}^nC_n + 2 . {}^nC_{n-1} . x + 3 . {}^nC_{n-2} . x^2 + . . . + (n+1) . {}^nC_0 . x^n $$

$$ \equiv (1 + x)^n + x . n(1 + x)^{n-1} $$

Putting x = 1, this becomes:

$$ {}^nC_n + 2 . {}^nC_{n-1} + 3 . {}^nC_{n-2} + . . . + (n+1) . {}^nC_0 $$

$$ = 2^{n-1} . (n + 2) $$

35. Find the sum of the first n terms of the series:

$$1 \cdot 2 + 2 \cdot 3 + 3 \cdot 4 + \ldots + n(n + 1) + .$$

Let S be the required sum, so that:

$$S = 1 \cdot 2 + 2 \cdot 3 + 3 \cdot 4 + \ldots + n(n + 1)$$

The general term n(n + 1) is of second degree in n. Let us assume that S is a polynomial of one degree higher in n, i.e. that it is possible to find constants a, b, c, d for which:

$$1 \cdot 2 + 2 \cdot 3 + 3 \cdot 4 + \ldots + n(n + 1) \equiv an^3 + bn^2 + cn + d \qquad (1)$$

This assumption can be justified. Replace n by n + 1, then:

$$1 \cdot 2 + 2 \cdot 3 + 3 \cdot 4 + \ldots + (n + 1)(n + 2) \equiv a(n + 1)^3 + b(n + 1)^2 + c(n + 1) + d \qquad (2)$$

Subtracting (1) from (2):

$$(n + 1)(n + 2) \equiv a\left\{(n + 1)^3 - n^3\right\} + b\left\{(n + 1)^2 - n^2\right\} + c(n + 1 - n)$$

or:

$$n^2 + 3n + 2 \equiv a(3n^2 + 3n + 1) + b(2n + 1) + c$$

Equating corresponding coefficients of powers of n on both sides:

$$3a = 1 \quad \therefore \quad a = \frac{1}{3}$$

$$3a + 2b = 3 \quad \therefore \quad b = 1$$

$$a + b + c = 2 \quad \therefore \quad c = \frac{2}{3}$$

Now to find d. Let $n = 1$ in (1) so that:

$$1 . 2 = a + b + c + d$$

$$\therefore \quad d = 0$$

Therefore, substituting for a, b, c and d:

$$S = \frac{1}{3}n^3 + n^2 + \frac{2n}{3} = \frac{1}{3}n(n + 1)(n + 2)$$

36. If $(5x - 6)^2 \geqslant x^4$, find the ranges of possible values of x.

(Ox)

The inequality is equivalent to:

$$x^4 - (5x - 6)^2 \leqslant 0$$

i.e. $$\{x^2 + (5x - 6)\}\{x^2 - (5x - 6)\} \leqslant 0$$

$$(x + 6)(x - 1)(x - 3)(x - 2) \leqslant 0$$

It is convenient to rearrange the factors in order of size:

$$(x - 3)(x - 2)(x - 1)\{x - (- 6)\} \leqslant 0 \qquad (1)$$

Clearly (1) will be satisfied if any of the four factors are zero, i.e. if $x = 3, 2, 1$ or -6. Apart from these values of x, (1) will be true only when there are an **odd** number of negative factors, either one or three.

For instance when $x > 3$, all factors, and hence their product are positive, so (1) is not satisfied.

When $3 > x > 2$, the signs of the four factors are in order $(-)(+)(+)(+)$, with a negative product, so (1) is satisfied.

Again, when $1 > x > -6$ the signs of the factors are $(-)(-)(-)(+)$ so the product is negative, thus satisfying (1).

And so the ranges of possible values of x are
$3 \geqslant x \geqslant 2$, $1 \geqslant x \geqslant -6$.

37. If the roots of the equation $x^3 - ax^2 + bx - c = 0$ are α, β, γ, find:

(a) The equation whose roots are:

$$2\alpha + \beta + \gamma \qquad 2\beta + \gamma + \alpha \qquad 2\gamma + \alpha + \beta$$

(b) The equation whose roots are:

$$\alpha^2 - \beta\gamma \qquad \beta^2 - \gamma\alpha \qquad \gamma^2 - \alpha\beta$$

(a) From the formulae connecting roots with coefficients a, b, c:

$$\alpha + \beta + \gamma = a \qquad (1)$$

Let:

$$2\alpha + \beta + \gamma = \alpha_1 \qquad 2\beta + \gamma + \alpha = \beta_1 \qquad 2\gamma + \alpha + \beta = \gamma_1$$

so that from (1):

$$\alpha + a = \alpha_1 \qquad \beta + a = \beta_1 \qquad \gamma + a = \gamma_1$$

Thus: $\alpha = \alpha_1 - a \qquad \beta = \beta_1 - a$

$$\gamma = \gamma_1 - a \qquad (2)$$

Now replace x by y - a, then the cubic becomes:

$$(y - a)^3 - a(y - a)^2 + b(y - a) - c = 0$$

Regarding this as a cubic in y instead of x, it follows from (2) that its roots are $\alpha_1 \beta_1 \gamma_1$, so it is the required cubic. On simplification it becomes:

$$y^3 - 4ay^2 + (5a^2 + b)y - (2a^3 + ab + c) = 0$$

(b) Let:

$$\alpha_1 = \alpha^2 - \beta\gamma \quad \beta_1 = \beta^2 - \gamma\alpha \quad \gamma_1 = \gamma^2 - \alpha\beta$$

Then:

$$\alpha_1 = \alpha^2 - \frac{\alpha\beta\gamma}{\alpha} = \alpha^2 - \frac{c}{\alpha} \text{ since } \alpha\beta\gamma = c$$

Similarly:

$$\beta_1 = \beta^2 - \frac{c}{\beta} \quad \text{and} \quad \gamma_1 = \gamma^2 - \frac{c}{\gamma}$$

This suggests taking:

$$y = x^2 - \frac{c}{x} \tag{3}$$

then finding an equation for y by eliminating x.

From (3):

$$x^3 - xy - c = 0$$

also:
$$x^3 - ax^2 + bx - c = 0$$

Subtracting and dividing across by x we get:

$$ax = b + y \quad \text{so:} \quad x = \frac{b + y}{a}$$

Substituting for x in:

$$x^3 - xy - c = 0$$

$$\left(\frac{b + y}{a}\right)^3 - y\left(\frac{b + y}{a}\right) - c = 0$$

which becomes, after reducing:

$$y^3 + y^2(3b - a^2) + by(3b - a^2) + b^3 - ca^3 = 0$$

whose roots, from (3) will be α_1, β_1, γ_1 as required.

38. Given that $p = N^2 + x$, where x is small compared with p, so that N is a close approximation to $\sqrt{(p)}$, expand:

$$\frac{N}{\sqrt{(p)}} \cdot \frac{N^2 + 3p}{3N^2 + p}$$

in ascending powers of $\frac{x}{p}$ as far as the term in:

$$\left(\frac{x}{p}\right)^3$$

Deduce that, neglecting terms of the order:

$$\frac{x^3\sqrt{(p)}}{p^3} \qquad \sqrt{(p)} = N \cdot \frac{N^2 + 3p}{3N^2 + p}$$

Use this result to obtain an approximation to the value of $\sqrt{(11)}$.

(Ox)

Substituting $p - x$ for N^2, $\frac{N}{\sqrt{(p)}} \cdot \frac{N^2 + 3p}{3N^2 + p}$ becomes:

$$\frac{\sqrt{(p - x)}}{\sqrt{(p)}} \cdot \frac{4p - x}{4p - 3x} = \left(1 - \frac{x}{p}\right)^{\frac{1}{2}}\left(1 - \frac{x}{4p}\right)\left(1 - \frac{3x}{4p}\right)^{-1}$$

expanding: $\left(1 - \frac{x}{p}\right)^{\frac{1}{2}}$ and $\left(1 - \frac{3x}{4p}\right)^{-1}$

by the binomial theorem this becomes:

$$\left(1 - \frac{x}{2p} - \frac{x^2}{8p^2} - \frac{x^3}{16p^3} - \ldots\right)\left(1 - \frac{x}{4p}\right)\left(1 + \frac{3x}{4p} + \frac{9x^2}{16p^2} + \frac{27x^3}{64p^3} + \ldots\right)$$

The coefficient of $\frac{x}{p}$ in this is:

$$\frac{3}{4} - \frac{1}{4} - \frac{1}{2} = 0$$

The coefficient of $\frac{x^2}{p^2}$ is:

$$-\frac{1}{8} + \frac{9}{16} + \frac{1}{8} - \frac{3}{16} - \frac{3}{8} = 0$$

The coefficient of $\frac{x^3}{p^3}$ is:

$$\frac{27}{64} - \frac{1}{16} - \frac{9}{64} - \frac{9}{32} + \frac{1}{32} + \frac{3}{32} - \frac{3}{32} = -\frac{1}{32}$$

Thus expansion is:

$$1 - \frac{1}{32}\left(\frac{x}{p}\right)^3 + \ldots \text{ higher powers of} \left(\frac{x}{p}\right)$$

Multiplying across by $\sqrt{(p)}$ we see that:

$$\sqrt{(p)} - \frac{1}{32} \cdot \sqrt{(p)}\left(\frac{x}{p}\right)^3 + \ldots = N \cdot \frac{N^2 + 3p}{3N^2 + p}$$

Hence if terms of order $\sqrt{(p)}\left(\frac{x}{p}\right)^3$ can be neglected:

$$\sqrt{(p)} = N \cdot \frac{N^2 + 3p}{3N^2 + p}$$

To find an approximation to $\sqrt{(11)}$, note that:

$$11 + \frac{1}{9} = \frac{100}{9} = \left(\frac{10}{3}\right)^2$$

Thus take:

$$N = \frac{10}{3}, \quad x = -\frac{1}{9}, \quad p = 11$$

so that:

$$\frac{x}{p} = -\frac{1}{99}$$

This fulfils the required condition of smallness.

Therefore an approximation to $\sqrt{(11)}$ is:

$$\frac{10}{3} \times \frac{11\frac{1}{9} + 33}{33\frac{1}{3} + 11} = \frac{10}{3} \times \frac{397}{399} = \frac{3970}{1197}$$

$$= 3{\cdot}316625$$

39. Prove that if a, b, c are the roots of the cubic equation:

$$t^3 - 3dt - p = 0$$

then:

$$(x^2 + ax + d)(x^2 + bx + d)(x^2 + cx + d) \equiv x^6 + px^3 + d^3$$

Express $x^6 - 20x^3 + 343$ as the product of three quadratic factors.

(Ox)

Let $x^2 + d = y$, so that:

$$(x^2 + ax + d)(x^2 + bx + d)(x^2 + cx + d) =$$

$$(y + ax)(y + bx)(y + cx)$$

$$= y^3 + y^2x(a + b + c) + yx^2(ab + bc + ca) + abcx^3 \quad (1)$$

Since a, b, c are the roots of $t^3 - 3d - p = 0$, we know that $a + b + c = 0$, $ab + bc + ca = -3d$, and $abc = p$.

Hence (1) becomes:

$$y^3 - 3dyx^2 + px^3$$

or: $$(x^2 + d)^3 - 3dx^2(x^2 + d) + px^3$$

This reduces to:

$$x^6 + px^3 + d^3 \qquad \text{i.e.}$$

$$(x^2 + ax + d)(x^2 + bx + d)(x^2 + cx + d) = x^6 + px^3 + d^3$$

Now to resolve $x^6 - 20x^3 + 343$ into three quadratic factors. Comparing it with $x^6 + px^3 + d^3$, we see that:

$$p = -20, \; d^3 = 343, \text{ so that } d = 7$$

Hence, here, a, b and c will be the roots of:

$$t^3 - 21t + 20 = 0$$

The remainder theorem shows that $t = 1$ is one root. Thus, dividing $t^3 - 21t + 20$ by $t - 1$, or otherwise, the other two roots are found to be 4 and -5.

$\therefore$ $a = 1$, $b = 4$, $c = -5$ and the required three quadratic factors of $x^6 - 20x^3 + 343$ are:

$$(x^2 + x + 7)(x^2 + 4x + 7)(x^2 - 5x + 7)$$

40. If α, β, γ are the roots of the equation $x^3 + ax + b = 0$, prove that:

$$\frac{\alpha^2 + \beta^2 + \gamma^2}{2} \cdot \frac{\alpha^3 + \beta^3 + \gamma^3}{3} = \frac{\alpha^5 + \beta^5 + \gamma^5}{5}$$

From the formulae connecting the roots and coefficients of a cubic:

$$\alpha + \beta + \gamma = 0, \; \alpha\beta + \beta\gamma + \gamma\alpha = a, \; \alpha\beta\gamma = -b$$

and so:

$$(1 - \alpha x)(1 - \beta x)(1 - \gamma x) \equiv 1 + ax^2 + bx^3$$

Taking logarithms to base e of both sides:

$$\log(1 - \alpha x) + \log(1 - \beta x) + \log(1 - \gamma x) \equiv$$

$$\log(1 + ax^2 + bx^3) \qquad (1)$$

Now when u is numerically less than 1, $\log_e(1 + u)$ may be expanded in the form of a series:

$$u - \frac{u^2}{2} + \frac{u^3}{3} - \frac{u^4}{4} + \ldots$$

Thus when x is small enough to make αx, βx, γx and $ax^2 + bx^3$ all numerically less than 1, expanding both sides of (1) it becomes:

$$-\alpha x - \frac{\alpha^2 x^2}{2} - \frac{\alpha^3 x^3}{3} - \ldots$$

$$-\beta x - \frac{\beta^2 x^2}{2} - \frac{\beta^3 x^3}{3} - \ldots$$

$$-\gamma x - \frac{\gamma^2 x^2}{2} - \frac{\gamma^3 x^3}{3} - \ldots$$

$$\equiv (ax^2 + bx^3) - \frac{(ax^2 + bx^3)^2}{2} + \frac{(ax^2 + bx^3)^3}{3} - \ldots (2)$$

It is convenient to denote, for short, $\alpha^2 + \beta^2 + \gamma^2$ by $\Sigma\alpha^2$, $\alpha^3 + \beta^3 + \gamma^3$ by $\Sigma\alpha^3$, and so on, Σ meaning summation.

Equating the coefficients of corresponding powers of x on both sides of (2):

$$\frac{\Sigma\alpha^2}{2} = -a, \quad \frac{\Sigma\alpha^3}{3} = -b, \quad \frac{\Sigma\alpha^4}{4} = \frac{a^2}{2}, \quad \frac{\Sigma\alpha^5}{5} = ab \qquad (3)$$

It follows from (3) that:

$$\frac{\Sigma\alpha^2}{2} \cdot \frac{\Sigma\alpha^3}{3} = (-a)(-b) = ab = \frac{\Sigma\alpha^5}{5}$$

which was to be proved.

41. Two functions f(x) and g(x) are said to commute if $f(g(x)) \equiv g(f(x))$. If f(x) and g(x) are quadratic and cubic polynomials, and $f(x) \equiv x^2 + a$, find the possible values of a if there is a g(x) that commutes with f(x), and find g(x) in each case.

(Ox)

Let $g(x) \equiv bx^3 + cx^2 + dx + e$, then:

$$f(bx^3 + cx^2 + dx + e) \equiv g(x^2 + a)$$

$$\therefore (bx^3 + cx^2 + dx + e)^2 + a \equiv b(x^2 + a)^3 + c(x^2 + a)^2$$

$$+ d(x^2 + a) + e \qquad (1)$$

Equating coefficients of x^6 gives $b^2 = b$, i.e. $b = 1$, the value $b = 0$ being inadmissible as g(x) is a cubic.

Equating coefficients of x^5, we get c = 0, so (1) is reduced to:

$$(x^3 + dx + e)^2 + a \equiv (x^2 + a)^3 + d(x^2 + a) + e \qquad (2)$$

Equating coefficients of x^4, x^3 etc on both sides:

$$2d = 3a, \quad 2e = 0, \quad \text{i.e.} \quad e = 0 \qquad (3)$$

$$d^2 = 3a^2 + d \qquad (4)$$

$$a = a^3 + da \qquad (5)$$

Solving (3) and (4) we see that a = 0, -2, d = 0, -3 which satisfy (5).

Therefore the possible values of a are 0 and -2, giving x^3 and $x^3 - 3x$ as the corresponding values of g(x).

42. Find:

$$\sum_{r=1}^{n} \frac{1}{(2r - 1)(2r + 1)(2r + 3)}$$

and deduce the value of:

$$\sum_{r=1}^{\infty} \frac{1}{(2r - 1)(2r + 1)(2r + 3)}$$

It is necessary first to express the general term:

$$\frac{1}{(2r - 1)(2r + 1)(2r + 3)}$$

in partial fractions.

Let:

$$\frac{1}{(2r-1)(2r+1)(2r+3)} \equiv \frac{A}{2r-1} + \frac{B}{2r+1} + \frac{C}{2r+3} \qquad (1)$$

where A, B, C are numbers to be determined.

Multiply both sides by $2r - 1$, and then let $r = \frac{1}{2}$ so that $2r - 1 = 0$. This shows that:

$$\frac{1}{(1+1)(1+3)} = A \quad \therefore \quad A = \frac{1}{8}$$

Multiplying similarly both sides by $2r + 1$ and $2r + 3$ in succession:

$$B = \frac{1}{(-1-1)(-1+3)} = -\frac{1}{4}$$

$$C = \frac{1}{(-3-1)(-3+1)} = \frac{1}{8}$$

Therefore from (1):

$$\frac{1}{(2r-1)(2r+1)(2r+3)} = \frac{1}{8}\left\{\frac{1}{2r-1} - \frac{2}{2r+1} + \frac{1}{2r+3}\right\}$$

Replacing r by the values 1, 2, 3 . . . n in succession:

$$\frac{1}{1 \cdot 3 \cdot 5} = \frac{1}{8}\left\{\frac{1}{1} - \frac{2}{3} + \frac{1}{5}\right\}$$

$$\frac{1}{3 \cdot 5 \cdot 7} = \frac{1}{8}\left\{\frac{1}{3} - \frac{2}{5} + \frac{1}{7}\right\}$$

$$\frac{1}{5 \cdot 7 \cdot 9} = \frac{1}{8}\left\{\frac{1}{5} - \frac{2}{7} + \frac{1}{9}\right\}$$

. . . .

. . . .

. . . .

$$\frac{1}{(2n-3)(2n-1)(2n+1)} = \frac{1}{8}\left\{\frac{1}{2n-3} - \frac{2}{2n-1} + \frac{1}{2n+1}\right\}$$

$$\frac{1}{(2n-1)(2n+1)(2n+3)} = \frac{1}{8}\left\{\frac{1}{2n-1} - \frac{2}{2n+1} + \frac{1}{2n+3}\right\}$$

Adding up, a number of terms on the right cancel, except at the beginning and end, so that:

$$\sum_{r=1}^{n} \frac{1}{(2r-1)(2r+1)(2r+3)}$$

$$= \frac{1}{8}\left\{1 - \frac{2}{3} + \frac{1}{3} + \frac{1}{2n+1} - \frac{2}{2n+1} + \frac{1}{2n+3}\right\} \qquad (2)$$

$$= \frac{1}{12} - \frac{1}{4(2n+1)(2n+3)}$$

To deduce the value of:

$$\sum_{r=1}^{\infty} \frac{1}{(2r-1)(2r+1)(2r+3)}$$

let $n \longrightarrow \infty$ in (2). Then:

$$\frac{1}{4(2n+1)(2n+3)} \longrightarrow 0$$

so that:

$$\frac{1}{12} - \frac{1}{4(2n+1)(2n+3)} \longrightarrow \frac{1}{12}$$

In other words:

$$\sum_{r=1}^{\infty} \frac{1}{(2r-1)(2r+1)(2r+3)} = \frac{1}{12}$$

43. Expand:

$$\frac{2 - 3x - 8x^2}{(1 - 2x)^2 . (1 - 3x)}$$

in a series of ascending powers of x as far as the term in x^3. State for what range of values of x the expansion is valid.

First resolve the given expression into partial fractions, which will have the form:

$$\frac{A}{(1 - 2x)^2} + \frac{B}{1 - 2x} + \frac{C}{1 - 3x}$$

A, B, C being numbers to be determined. By equating corresponding coefficients of x on both sides, or otherwise, it is found that:

$$A = 3, \quad B = -2, \quad C = 1$$

Thus:

$$\frac{2 - 3x - 8x^2}{(1 - 2x)^2(1 - 3x)} \equiv \frac{3}{(1 - 2x)^2} - \frac{2}{(1 - 2x)} + \frac{1}{(1 - 3x)}$$

$$\equiv 3(1 - 2x)^{-2} - 2(1 - 2x)^{-1}$$

$$+ (1 - 3x)^{-1} \qquad (1)$$

Now if a is numerically less than 1, i.e. if $-1 < a < +1$, we can expand $(1 - a)^{-2}$ by the Binomial Theorem in infinite series form:

$$1 - 2(-a) + \frac{(-2)(-3)(-a)^2}{2!} + \ldots = 1 + 2a + 3a^2 + 4a^3 + \ldots$$

And, for the same range of values of a, similarly:

$$(1 - a)^{-1} = 1 + (-1)(-a) + \frac{(-1)(-2)(-a)^2}{2!} + \ldots$$

$$= 1 + a + a^2 + a^3 + \ldots$$

Thus $(1 - 2x)^{-2}$, $(1 - 2x)^{-1}$ and $(1 - 3x)^{-1}$ may all be expanded by the Binomial Theorem if both:

$$-1 < 2x < +1 \quad \text{and} \quad -1 < 3x < +1$$

These inequalities are both true when:

$$-\frac{1}{3} < x < +\frac{1}{3}$$

And so, from (1) when:

$$-\frac{1}{3} < x < +\frac{1}{3}$$

$$\frac{2 - 3x - 8x^2}{(1 - 2x)^2 . (1 - 3x)}$$

$$\equiv 3(1 + 2 . (2x) + 3 . (2x)^2 + 4 . (2x)^3 + \ldots)$$

$$- 2(1 + (2x) + (2x)^2 + (2x)^3 + \ldots)$$

$$+ (1 + 3x + (3x)^2 + (3x)^3 + \ldots)$$

$$\equiv 2 + 11x + 37x^2 + 107x^3 + \ldots$$

44. (a) If x and y vary so that $x \geqslant 0$, $y \geqslant 0$, and that both:

$$3x + 2y \leqslant 40$$

and $$5x + 7y \leqslant 85$$

find the maximum value of $4x + 3y$.

(b) If x and y vary so that $x \geqslant 0$, $y \geqslant 0$, and that both:

$$3x + y \geqslant 15$$

and $$4x + 3y \geqslant 30$$

find the minimum value of $2x + y$.

(a) Let $4x + 3y \equiv p(3x + 2y) + q(5x + 7y)$ where p and q are constants to be found. Equating coefficients of x and of y on both sides, we get:

$$3p + 5q = 4 \qquad 2p + 7q = 3$$

so that:

$$p = \frac{13}{11} \qquad q = \frac{1}{11}$$

which, note, are both **positive**. Therefore:

$$4x + 3y \equiv \frac{13}{11}(3x + 2y) + \frac{1}{11}(5x + 7y)$$

It follows that when both $3x + 2y$ and $5x + 7y$ have their maximum values so will $4x + 3y$. Thus the maximum value of $4x + 3y$ occurs when $3x + 2y = 40$, and $5x + 7y = 85$, i.e. when $x = 10$, $y = 5$. Its value is 55.

(b) As in (a) let $2x + y \equiv p(3x + y) + q(4x + 3y)$. Equating coefficients of x and y:

$$3p + 4q = 2 \qquad p + 3q = 1$$

from which:

$$p = \frac{2}{5} \qquad q = \frac{1}{5}$$

again both positive. And so:

$$2x + y \equiv \frac{2}{5}(3x + y) + \frac{1}{5}(4x + 3y)$$

whose minimum value occurs when $3x + y$ and $4x + 3y$ both attain their minimum values. Therefore $2x + y$ has its minimum value when $3x + y = 15$, $4x + 3y = 30$, i.e. when $x = 3$, $y = 6$.

This minimum value is 12.

45. If $z = x + iy$ and $\bar{z} = x - iy$, show that the equation:

$$z\bar{z} - a\bar{z} - \bar{a}z + c = 0$$

where a is a complex constant, and c is a real constant, represents a circle in the Argand diagram.

Find the complex number which corresponds to the centre, and show that the square of the radius is $a \cdot \bar{a} - c$.

Find the condition that the origin should lie inside the circle. (Ox)

Let $a = p + iq$, where p and q are real, then $\bar{a} = p - iq$. Thus $z\bar{z} - a\bar{z} - \bar{a}z + c = 0$ is equivalent to:

$$(x + iy)(x - iy) - (p + iq)(x - iy) - (p - iq)(x + iy) + c = 0$$

i.e. $$x^2 + y^2 - 2px - 2qy + c = 0$$

This represents a circle in the Argand diagram.

The co-ordinates of its centre are p and q, so the complex number which corresponds to it is p + iq, or a.

The square of the radius of the circle is:

$$p^2 + q^2 - c = (p + iq)(p - iq) - c$$

$$= a \cdot \bar{a} - c$$

The square of the distance between the centre of the circle and the origin is:

$$p^2 + q^2 = a \cdot \bar{a}$$

Therefore if:

$$a\bar{a} < a\bar{a} - c, \text{ i.e. if } c < 0$$

the origin will be inside the circle.

PART II - NOTES

1. (a) More generally, whenever a function y, of x, can be expressed as the sum of partial fractions such as, for instance:

$$y = \frac{3}{x - 4} + \frac{5}{x + 3} - \frac{6}{(x + 2)^2}$$

use this method to find: $\frac{d^n y}{dx^n}$

For the n th derivative of each partial fraction:

$$3(x - 4)^{-1}, \quad 5(x + 3)^{-1}, \quad -6(x + 2)^{-2}$$

is found as in 1.(a).

(b) This technique, putting $a \cos \theta + b \sin \theta$ in the forms $r \sin (\theta + \alpha)$
or, $r \cos (\theta + \beta)$,
should be remembered; it is often needed.

2. Used here is the fact that if a function f(x) is positive when $a \leqslant x \leqslant b$, except possibly for a limited number of values of x for which f(x) = 0, then:

$$\int_a^b f(x) \, . \, dx > 0$$

This may be proved graphically or otherwise, and is often helpful. For example, the function 1 - cos x satisfies the required conditions in any interval, and so if $u > 0$:

$$\int_0^u (1 - \cos x) \, . \, dx > 0$$

i.e. $u - \sin u > 0$, so $u > \sin u$.

Repeating the procedure now with the function $x - \sin x$, we get:

$$\int_0^u (x - \sin x) \, . \, dx > 0$$

i.e. $$\cos u > 1 - \frac{u^2}{2}$$

A further repetition now with the function:

$$\cos x + \frac{x^2}{2} - 1$$

gives: $$\sin u > u - \frac{u^3}{3!}$$

Clearly it can be repeated indefinitely.

3. (a) In general if, (as here), the integrand contains **one** radical only, for example $\sqrt{(5x - 2)}$, the substitution $5x - 2 = u^2$ will work.

(b) A similar device can be used for a number of other integrals. For example, to find:

$$\int \frac{x^5 dx}{(x^3 + 4)^3}$$

write the integrand as:

$$\frac{x^3 \, . \, x^2 dx}{(x^3 + 4)^3}$$

then make the substitution $x^3 + 4 = u$. This transforms the integral into:

$$\int \frac{(u - 4)du}{3u^3} = \frac{1}{3}\left(-\frac{1}{u} + \frac{2}{u^2}\right)$$

(c) Usually when $\cos^2 \theta$ or $\sin^2 \theta$ occur in an integrand they must be replaced, respectively, by $\frac{1}{2}(1 + \cos 2\theta)$ and $\frac{1}{2}(1 - \cos 2\theta)$. This was done here. This is not a general rule, however, for example:

$$\int \cos^2 \theta \sin \theta \, . \, d\theta$$

The work indicates how to find an integral like:

$$\int \sin ax \, . \, \cos bx \, . \, dx$$

where $a > b$ say, by writing it as:

$$\frac{1}{2} \int \{\sin ((a + b)x) + \sin ((a - b)x)\} \, dx$$

$$= -\frac{1}{2} \left(\frac{\cos((a + b)x)}{a + b} + \frac{\cos((a - b)x)}{a - b}\right)$$

The integrals $\int \cos ax \, . \, \cos bx \, . \, dx$,

and $\int \sin ax \, . \, \sin bx \, . \, dx$ are found similarly.

4. Most questions of this nature can be done like this by performing two integrations, or sometimes only one, depending on what has been asked. After each integration a constant of integration has to be found by using some of the given conditions, as was done here.

In the case of a body moving along a straight line having described a distance S in time t, with a velocity v at that instant, be ready to recognize problems which are solvable by this method. Here:

$$\frac{d^2S}{dt^2}$$

is the acceleration, and:

$$\frac{dS}{dt}$$

is the velocity v. Remember, too, that another expression for the acceleration is:

$$v \frac{dv}{dS}$$

This is often needed.

5. The problem here is to find two unknown functions of x, namely y and z. No general advice can usefully be given to help the student in all such cases. He should, however, as was done here, try to manipulate the two given equations so as to obtain two other equations each containing only **one** unknown function, such as (1) and (2) here. Their solution, by performing an integration, will usually be straightforward.

6. (a) Follow this procedure in finding the equations of tangents and normals to curves given in Cartesian co-ordinates. The only thing to mention is that sometimes the equation of the given curve is more complicated than that here, which makes the expression for:

$$\frac{dy}{dx}$$

more cumbersome.

(b) Let two lines have gradients m_1 and m_2. Then if α is either of the angles between the lines:

$$\tan \alpha = \frac{m_1 - m_2}{1 + m_1 m_2}$$

If this is positive, it gives the acute angle between the lines, if it is negative it gives the obtuse angle between them.

7. The solution given here depends entirely on what may be called Test 1 for maxima and minima of a function $y = f(x)$. This states that if:

$$\frac{dy}{dx} = 0$$

when $x = a$, and its sign changes from being positive to negative there, then y has a maximum value when $x = a$. If:

$$\frac{dy}{dx} = 0$$

when $x = a$, but its sign changes from being negative to positive there, then y has a minimum value when $x = a$.

It was necessary during the working to differentiate functions such as:

$$e^{-x^2} \quad x^{x^2}$$

There is an alternative procedure in such cases which you may prefer. Let:

$$y = e^{-x^2} . x^{x^2}$$

Take logarithms of both sides to base e:

$$\log y = -x^2 + x^2 . \log x$$

Differentiate both sides with respect to x.

$$\frac{1}{y} . \frac{dy}{dx} = -x + 2x \log x$$

so that: $$\frac{dy}{dx} = y(-x + 2x \log x)$$

$$= e^{-x^2} . x^{x^2} . x(2 \log x - 1)$$

which is the same as found before.

8. Many, though not all, of the other reduction formulae met with are also proved by integrating by parts. It is the first method to try. Here is an instance. If:

$$I_n = \int_0^{\pi/2} \sin^n x \, . \, dx$$

to prove that:

$$I_n = \frac{n - 1}{n} \, . \, I_{n-2}$$

where n is a positive integer: ($\geqslant$ 2).

Consider:

$$I_n = \int_0^{\pi/2} \sin^n x \, . \, dx$$

$$= \int_0^{\pi/2} \sin^{n-1} x \, . \, \sin x \, . \, dx$$

$$= \int_0^{\pi/2} \sin^{n-1} x \, . \, \frac{d(-\cos x)}{dx} \, . \, dx$$

Integrate by parts:

$$= \Big[-\sin^{n-1} x \, . \, \cos x \Big]_0^{\pi/2}$$

$$+ (n - 1) \int_0^{\pi/2} \sin^{n-2} x \, . \, \cos^2 x \, . \, dx$$

$$= (n - 1) . \int_0^{\pi/2} \sin^{n-2} x \, . \, (1 - \sin^2 x) \, . \, dx$$

$$= (n - 1)(I_{n-2} - I_n)$$

i.e. $$I_n = \frac{(n - 1)}{n} \, . \, I_{n-2}$$

9. (a) Illustrates again the advantage of replacing:

$$\cos^2 x \text{ by } \frac{1 + \cos 2x}{2}$$

when it occurs in an integrand, though there will be found exceptions to this (see the note on 3.(c)). With regard to $\int x \cos 2x \, . \, dx$ it is possible to do even the general case $\int x \cos kx . dx$ without writing down any substitution of u. For, integrating by parts, mentally, thinking of kx as u:

$$\int x \cos kx \, . \, dx = x \, . \, \frac{\sin kx}{k} - \int \frac{\sin kx}{k} . dx$$

$$= \frac{x \sin kx}{k} + \frac{\cos kx}{k^2}$$

(b) The general case:

$$\int \sqrt{\left(\frac{a - x}{b - x}\right)} \, . \, dx$$

is more difficult, but we start in the same way by rationalizing the numerator. The integral is now:

$$\int \frac{(a - x)dx}{\sqrt{((a - x)(b - x))}} = \int \frac{(a - x)dx}{\sqrt{(x^2 - x(a + b) + ab)}}$$

Making the usual substitution here, namely:

$$x - \frac{a + b}{2} = u$$

the integral becomes:

$$\int \frac{\left(\frac{a - b}{2} - u\right) du}{\sqrt{\left(u^2 - \left(\frac{a - b}{2}\right)^2\right)}}$$

which consists of two standard integrals in u.

10. (a) The following facts are used here, and it is necessary to know them for this and similar questions.

When the equation of a straight line is, as here, expressed in the form $x \cos \alpha + y \sin \alpha - p = 0$, if Q is the point $(x_1 y_1)$, then $x_1 \cos \alpha + y_1 \sin \alpha - p < 0$ when Q is on the same side of the line as O. The length of the perpendicular from Q to the line is $p - x_1 \cos \alpha - y_1 \sin \alpha$. Similarly if Q is on the opposite side of the line from 0, $x_1 \cos \alpha + y_1 \sin \alpha - p > 0$, and is, now, the length of the perpendicular from OQ to the line.

(b) There are other problems of this kind where a similar minimum value is sought. One, for instance, is to find the length of the longest straight plank which can be brought, horizontally, round a corner where two corridors, of widths a and b, intersect at right angles. Here the function of α to be minimized is:

$\frac{a}{\sin \alpha} + \frac{b}{\cos \alpha}$ Its minimum value is: $\left(a^{2/3} + b^{2/3}\right)^{3/2}$

and this will be the length of the longest plank fulfilling the conditions.

11. (a) A straightforward application of Test 1 for maxima and minima.

(b) A point of inflexion on a curve is one where the gradient:

$$\frac{dy}{dx}$$

has a maximum or minimum value. Therefore two requirements must be satisfied at a point of inflexion, namely:

(1) $$\frac{d}{dx} \cdot \left(\frac{dy}{dx}\right) \quad \text{or} \quad \frac{d^2y}{dx^2} = 0$$

(2) $\frac{d^2y}{dx^2}$ must change sign there.

12. The cases where $x \geqslant 1$ and $x < 0$ present no difficulty, and, as was shown, one is deducible from the other. In dealing with the third case where $0 < x < 1$ it was necessary to divide the range of integration:

$$\int_0^1 \text{ into two parts: } \int_x^1 \text{ and } \int_0^x$$

The reason was that, in the first, $|x - t| = t - x$, whereas, in the second, $|x - t| = x - t$.

13. Notice that the integrals appearing in (a) and (b) are both particular cases. Thus:

$$\int x^n \log x \,.\, dx = \frac{x^{n+1} \,.\, \log x}{n + 1} - \frac{x^{n+1}}{(n + 1)^2}$$

on integrating by parts. Now consider:

$$\int e^{x^n} \,.\, x^{n-1} \,.\, dx$$

The substitution $u = x^n$ transforms it into:

$$\frac{1}{n}\int e^u \,.\, du = \frac{e^u}{n} = \frac{e^{x^n}}{n}$$

The formula used in (c) evaluates many definite integrals which appear formidable. It is easily proved by letting $a - x = u$.

14. This is included to provide some practice in using (a) the formulae for finding the centroid of a plane area, and (b) the formula for finding the centroid of a solid of revolution.

15. (a) Three of the basic formulae, with which the student has to be familiar, are necessary here. They are:

(1) $$\frac{d \sin^{-1} u}{du} = \frac{1}{\sqrt{(1 - u)^2}}$$

one of the standard formulae in differentiation.

(2) The 'quotient formula' for differentiating a quotient v/w.

(3) The 'function of a function' formula, stated during the working of the question.

(b) Actually it is just as easy to deal with the general case in which $y = F(x)^{f(x)}$, F(x) and f(x) being any pair of differentiable functions. To find:

$$\frac{dy}{dx}$$

Since:

$$F(x) = e^{\log F(x)}$$

it follows that:

$$y = e^{f(x) \,.\, \log F(x)} = e^u \quad \text{where:}$$

$$u = f(x) \,.\, \log F(x)$$

Denote: $\frac{df(x)}{dx}$ by $f'(x)$, and: $\frac{dF(x)}{dx}$ by $F'(x)$

Then: $$\frac{dy}{dx} = \frac{dy}{du} \cdot \frac{du}{dx}$$

$$= e^u(f^1(x) \log F(x) + f(x) \cdot \frac{F^1(x)}{F(x)})$$

$$= F(x)^{f(x)}(f^1(x) \log F(x) + f(x) \cdot \frac{F^1(x)}{F(x)})$$

A second method, which many prefer, is to take logarithms on both sides of the equation:

$$y = F(x)^{f(x)}$$

to base e, and then to differentiate with respect to x. See the note on example 7. Work it by this second method, too, for practice.

16. (a) Generally, when given the equation of a curve and asked to sketch, or trace, it, here are the main things to look for:

(1) Does it pass through the point (0, 0) or the origin?

(2) Where does it cut the x and y axes?

(3) Is it symmetrical about the x axis, i.e. do only even powers of y appear in the equation?

(4) Similarly for the y axis.

(5) Replace x, y by -x, -y. If this leaves the curve unaltered it is symmetrical about the origin, so its shape in the first quadrant is repeated in the third, and in the second quadrant is repeated in the fourth.

(6) Has the curve any maxima or minima?

(7) Has it any asymptotes?

(8) What happens when:

$$x \longrightarrow \begin{cases} +\infty \\ -\infty \end{cases}, \text{ when: } y \longrightarrow \begin{cases} +\infty \\ -\infty \end{cases} ?$$

(b) If we took, instead, $y^2 = x^2(a - x)$, where a is any positive number, the centroid would be found to be:

$$\left(\frac{4a}{7}, 0\right)$$

Observe that the integrals in (a) and (b) were found by applying the general rule stated in the note on 3.(a).

17. (a) As a general rule whenever an integral of type:

$$\int \frac{(x + p)dx}{\sqrt{(x^2 + 2ax + b)}}$$

turns up, make the substitution $x + a = u$, so that the transformed integral is:

$$\int \frac{(u + p - a)du}{\sqrt{(u^2 + b - a^2)}}$$

Now according to whether $b - a^2$ is positive or negative:

$$\sqrt{(u^2 + b - a^2)}$$

takes the form:

$$\sqrt{(u^2 + k^2)} \quad \text{or:} \quad \sqrt{(u^2 - k^2)}$$

Whichever is the case, the integrals in u resulting from the substitution are of standard types, for:

$$\int \frac{du}{\sqrt{(u^2 \pm k^2)}} = \log (u + \sqrt{(u^2 \pm k^2)})$$

and:

$$\int \frac{udu}{\sqrt{(u^2 \pm k^2)}} = \sqrt{(u^2 \pm k^2)}$$

(b) This substitution will also work with integrals such as:

$$\int \frac{d\theta}{4 \sin^2 \theta + 3}$$

Since $\cos^2 \theta + \sin^2 \theta = 1$, the integral may be written as:

$$\int \frac{d\theta}{7 \sin^2 \theta + 3 \cos^2 \theta}$$

18. (a) Whenever, as here, x and y are expressed in terms of a variable parameter θ by equations of the form $x = f(\theta)$, $y = g(\theta)$, follow this same procedure in order to find:

$$\frac{d^2y}{dx^2}$$

(b) Now try the general case. If $x^2 + 2axy + y^2 = 1$, prove that:

$$\frac{d^2y}{dx^2} = \frac{(a^2 - 1)}{(ax + y)^3}$$

in the same way.

19. After finding the expression:

$$\frac{d^2u_n}{dx^2} = n(n-1)\{x(1-x)\}^{n-2} . (1-2x)^2$$

$$- 2n\{x(1-x)\}^{n-1}$$

a regrouping of terms was necessary to get it into the required form:

$$n(n-1)u_{n-2} - 2n(2n-1)u_{n-1}$$

There is no guidance which can usefully be given about how to do this. Several unsuccessful regroupings may have to be tried before discovering the correct one. Regarding:

$$v_n = \int_0^1 e^x \cdot u_n \cdot dx$$

its very form:

$$\int_0^1 u_n \cdot \frac{d(e^x)}{dx} \cdot dx$$

suggests integration by parts, and then a repetition of this. The same is true of w_n.

20. Note that h was eliminated, not r, because (1) is a **linear** equation in h, which makes it easy to do so, whereas it is a **quadratic** equation in r.

Suppose, generally, that V is a function of x and y, whose maximum (or minimum) value is sought. Suppose, too, that x and y are not independent, but satisfy an equation which is linear in, say, y. This equation enables us to eliminate y, and hence express V as a function of x alone. Now follow the same procedure as here. Other problems like this will be met with, whose solution can be found in this way.

21. Here, again, little guidance can be given. One has to try various manipulations of the three given equations. However, an examination of (a) reveals the fact that it contains only u and v, together with their first and second derivatives, thus it should be deducible from the equation $uv = a$, which it is.

The fact that $uv'' + u''v + 2u'v' = 0$ suggested multiplying (i) by v and (ii) by u, since, among others, it provided the terms $uv'' + vu''$, and $p(x)(u'v + v'u)$ which is zero. This is an encouragement to proceed, proving (b).

Again:

$$a\frac{dq(x)}{dx} = u'v'' + u''v'$$

got by differentiating (b), suggests multiplying (i) by v', (ii) by u' and adding, thus obtaining among other terms, $u'v'' + u''v'$.

22. This technique is effective in finding quickly the equations of the lines joining the origin O to the points where a given line and curve intersect. Suppose, for another instance, that the equations of the given line and curve are respectively:

(A) $2x + 5y = 8$, and:

(B) $x^2 + 3xy + 2y^2 - 4x - 5y - 1 = 0$

From (A):

$$\frac{2x + 5y}{8} = 1$$

This can be used to derive a homogeneous equation of the second degree from (B) in the following way. Consider the equation:

(C) $$x^2 + 3xy + 2y^2 - (4x + 5y)\left(\frac{2x + 5y}{8}\right) - \left(\frac{2x + 5y}{8}\right)^2 = 0$$

It clearly is satisfied by the points where the given line and curve intersect, say P and Q.

Furthermore (C) being of form:

$$ax^2 + 2hxy + by^2 = 0$$

represents a pair of lines through O, which must, therefore, be OP and OQ.

23. Many questions on the parabola or hyperbola (and also ellipse) are made easier to solve by using parameters, like t and T in this instance. As it is required so frequently, it will save time if you can remember the equation of the tangent to the parabola $y^2 = 4ax$ at the point 't' on it, namely:

$$y = \frac{x}{t} + at$$

It is, of course, essential to remember the equation of the line joining the points (x_1y_1) and (x_2y_2), i.e:

$$\frac{y - y_1}{y_1 - y_2} = \frac{x - x_1}{x_1 - x_2}$$

used here.

24. In drawing a plane diagram to represent a three dimensional object, it may help to emphasize the fact by representing lines such as CA, CN, DN, which are not visible, by broken lines.

On the line of intersection of two given planes there will usually be one particular point, like N in this case. Its special position in the diagram makes it easy to calculate the angle between the perpendiculars to the line of intersection drawn one in each plane, which will be the required angle between the planes. Look out for it every time.

In the case of two skew lines like VA and BC, there is one line which can be drawn perpendicular to and intersecting both of them. It is a question of finding it. If it intersects the two skew lines at P and Q then PQ will be the shortest distance between them.

25. This method can be employed in other questions. For instance, to find the locus of the middle points of parallel chords of the ellipse making an angle α, say with the x axis. Consider the same equation for r:

$$\frac{(x_1 + r\cos\alpha)^2}{a^2} + \frac{(y_1 + r\sin\alpha)^2}{b^2} = 1$$

If $A(x_1 y_1)$ is the middle point of the chord PQ (same diagram), then the two values of r obtained from this equation must be equal and opposite. And so their sum is zero, i.e:

$$\frac{x_1\cos\alpha}{a^2} + \frac{y_1\sin\alpha}{b^2} = 0$$

Replacing x_1 and y_1 by 'running' co-ordinates x and y gives the required locus:

$$\frac{x\cos\alpha}{a^2} + \frac{y\sin\alpha}{b^2} = 0$$

which is a diameter of the ellipse.

26. Suppose the co-ordinates x and y of a variable point P can be expressed in terms of a parameter t. Then there are two equations like:

(1) $x = f(t)$

(2) $y = g(t)$

The equation of the locus of P will be found by eliminating t between (1) and (2).

However, in the present question, the co-ordinates of P are expressible in terms of two parameters $t_1 + t_2$ and t_1t_2 by equations (5) and (6). A third equation, therefore, is needed in order to eliminate the two parameters. This is provided by equation (4), and the locus of P is then found.

27. This does not require much comment. If C is the centre of any sphere through P and Q, then CP = CQ, so that C must lie on a perpendicular bisector of PQ. Now all the perpendicular bisectors of PQ lie on the plane which bisects PQ at right angles. Hence this plane is the locus of the centres of all spheres through P and Q, which proves the statement numbered (1).

28. When, as here, x and y are expressed in terms of a parameter t by the equations $x = f(t)$, $y = g(t)$, remember that the equation of the tangent to the curve at the point 't' takes the form:

$$\frac{y - g(t)}{x - f(t)} = \frac{g'(t)}{f'(t)}$$

dashes denoting differentiation with respect to t.

You will notice that, as in example 26, it is necessary to eliminate two parameters from three equations in order to find the locus of (X, Y).

29. Any point on the ellipse can be taken as:

$$(\sqrt{(2)}\, p \cos \theta, \quad \sqrt{(2)}\, q \sin \theta)$$

With regard to equation (1), note that, though t_1 and t_2 are its roots, it is not necessary to calculate them separately. For only the symmetrical terms $t_1 + t_2$ and t_1t_2 are required in (2), and these can be written down from (1).

The same thing is true about u_1 and u_2, the roots of equation (4). Only $u_1 + u_2$ and u_1u_2 are required in (5), and they can be written down from (4).

30. Another case of eliminating two parameters from three equations to find a locus. After examining equations (1) to (3), you will notice that it is better to treat $t_1 + t_2$ and $t_1^2 + t_2^2$ as the quantities to be eliminated, because doing so shortens the work.

31. This problem is of a type not commonly met with. The procedure only works because of the special formation of the four equations. It, or a modification of it, is, however, occasionally used elsewhere, for instance in certain questions of elimination.

32. This question, too, is one of those original types about whose solution no general advice can be given. However, the main thing is to formulate equations (1) to (3); having done so, the rest follows without much difficulty.

33. Dealing with inequalities like this, avoid multiplying across by the denominator $x(x - 4)$, thus obtaining $(2x - 3)(2x - 1) \geqslant 3x(x - 4)$. This is true only on the unjustified assumption that $x(x - 4)$ is positive. It is safer to proceed by **subtraction** of terms. For instance, deal with the inequality:

$$\frac{3x - 1}{x - 5} > 2$$

by putting it in the form:

$$\frac{3x - 1}{x - 5} - 2 > 0, \text{ i.e. } \frac{x + 9}{x - 5} > 0$$

Using the method of example 33, this is seen to be true when either $x > 5$ or $x < -9$.

34. This is one of a great variety of questions which can be set on permutations and combinations. Nothing can usefully be said here which will apply to any large number of these. Familiarity with the properties of the binomial coefficients such as, for example:

$$^nC_0 + {}^nC_1 + {}^nC_2 + \ldots + {}^nC_n = 2^n$$

is necessary, and with textbook theorems in permutations and combinations.

35. The method here has a wide application. It will find the sum of the first n terms of any series whose general (n th) term is a given polynomial in n, say $n^3 + 3n^2 - 4n + 2$. Suppose that S is the sum of the first n terms of this series. Here the n th term is of the **third** degree in n. Now assume that S is a polynomial of **one** degree higher in n, i.e. that:

$$S \equiv an^4 + bn^3 + cn^2 + dn + e$$

where a, b, c, d, e are numbers.

It is possible to justify this assumption by finding suitable values for a, b, c, d, and e, using the method of this example. For practice, prove, similarly, that if the n th term of a series is n^3, the sum of its first n terms will be:

$$\frac{n^2(n+1)^2}{4}$$